KU-615-875

WITHDRAWN
FROM
CIRCULATION

Stockport Libraries

C2000003230202

Birdsong in a Time of Silence

Birdsong in a Time of Silence

Steven Lovatt

PARTICULAR BOOKS

an imprint of

PENGUIN BOOKS

PARTICULAR BOOKS

UK | USA | Canada | Ireland | Australia
India | New Zealand | South Africa

Penguin Books is part of the Penguin Random House group of companies whose
addresses can be found at global.penguinrandomhouse.com.

First published by Particular Books 2021

001

Text copyright © Steven Lovatt, 2021
Illustrations copyright © Katie Marland, 2021

The moral right of the author has been asserted

Set in 12.5/14.75pt Fournier MT Pro
Typeset by Jouve (UK), Milton Keynes
Printed and bound in Great Britain by Clays Ltd, Elcograf S.p.A.

The authorized representative in the EEA is Penguin Random House Ireland,
Morrison Chambers, 32 Nassau Street, Dublin D02 YH68

A CIP catalogue record for this book is available from the British Library

ISBN: 978–0–241–49300–7

www.greenpenguin.co.uk

MIX
Paper from
responsible sources
FSC C018179

Penguin Random House is committed to a
sustainable future for our business, our readers
and our planet. This book is made from Forest
Stewardship Council® certified paper.

To those imprisoned, who cannot hear birdsong.

Contents

The fuchsia tree

O what if the fowler my blackbird has taken?
The roses of dawn blossom over the sea.
Awaken, my blackbird, awaken, awaken!
And sing to me out of my red fuchsia tree!

Trad. Manx

House sparrow

1. The Strangest Spring

It's six in the morning and still dark, the 24th of March, 2020. I wake early, and, knowing the children will soon be up, make the sudden decision to steal half an hour's solitude in the park. From the dense latticework of trees and shrubs that clothe the wooded slope comes a constant scuttling through dead leaves, and a restless readjustment of twigs. The darkness is awake and vigilant; there's the warning *tik-tik* of an invisible robin from the bushes, and then the next second it appears on the path. Each individual movement of the bird, each wing-flick and pivot, is brisk and definite, yet the overall impression is one of nervousness and indecision. It leaps round once more on the spot, then flits back into the darkness. From close by comes a blast of song from a wren, its harsh trill like coarse twine zipping over a fly-wheel. The air is cool, not cold, and smells deliciously of earth and moss. With every moment, almost visibly, more light slips into the air as the sun creeps closer to the horizon.

There's a sudden disturbance from the deeper shade, and a blackbird comes careering out with a mad clatter, as though a fragment of the shadow itself had been flung away from some explosive centre and taken the form of the bird that now pauses in alert rest on the great arm of the beech tree below which I'm standing, pressed quiet against the trunk. Although I can see it only in silhouette, a cut-out in stiff black paper against the felty shade of the trees, it's evidently agitated. It flicks about the bough, dipping then raising its wings, and tilting its head all the while in response to something I can't sense. After a few seconds of this twitching the bird seems to experience some sort of inner resolution, and as the first beam of grey light wakes the colours of the tree it raises its head and lets out a quiet phrase of song. Spring has arrived.

The day before my early walk in the park, the prime minister ordered a shut-down of public life that would entirely change society as we'd known it. Travel was forbidden; schools closed and playground gates were locked; cafés and pubs were shuttered; and doctors' surgeries turned into fortresses, plastered with warning signs, only accessible by phone. A new animal, a microscopic animal, given the name Covid-19, was racing around the world faster than even the sleepless global media could track it. Although invisible, it spread fear as effectively as any dragon from a folk tale. Nobody knew where it was, but it took its tribute of anxiety and silence all the same. So it was that, by government decree, normal life was suspended.

Where I live, compliance was immediate and total. All traffic noise ceased, and you could hear litter scuffing down the empty streets. Paper rainbows began to appear in windows, painted as a token of hope by children kept indoors; but of the children themselves there was no sign. It felt less like a catastrophe than an aftermath, as if nine-tenths of the population had disappeared overnight. Groping for language to understand what was happening, both television pundits and ordinary people fell back on a wartime vocabulary of discipline and solidarity. Yet the strangeness was amplified tenfold by the difficulty of reconciling this 'lockdown' with the sudden coming of the most glorious spring that anyone could remember. After a long winter of gales and rain came still, warm days. Our spring bulbs emerged, magnolias brought out their menorahs, and even as the supermarket shelves were emptied of dried pasta, yeast and toilet roll, astonishing scents and colours were being finalized for display just an inch below the surface of the mild, wet earth.

But most of all, we began to notice the birdsong. A little tentative and sputtering at first, by the end of March it filled the air. Broadcast from aerials and hedge-tops, a rising choir of chirps, trills and warbles brought life to gardens and echoed off housefronts, shuttered shops and bland retail silos, seeming suddenly obtrusive with no motor traffic to smother it. Some bird calls are present all year around, and these are among the easiest to recognize. Everyone knows the crow's harsh croak, and one needn't live near the coast to hear the oily yelp of gulls. But with the awakening spring came other songs that were harder to place. One sounds a little like a five-pence coin being dropped over and again to skitter on a varnished table. Another resembles a fly-past of wind-chimes. Yet another bird makes a jingle of notes, like water from a fountain, but somehow feeble, as if the water pressure were too low. As lockdown continued, we became curious about these calls, and peered down from balconies or went out into the garden to see what had made them. And as the days passed, I was able to take advantage of my good fortune to live in a green, coastal town and within the bounds of the permitted daily exercise reawaken an interest in birds that had been pressed to the margins of my life by the responsibilities of parenthood and work.

I started watching birds when I was seven, by which time I'd already passed through various other intense but short-lived enthusiasms for cars, dinosaurs, spacecraft and warplanes. I don't really know why the bird phase stuck, but I probably owe it to my parents, both of whom encouraged my interest and soon became enthusiasts in their own right. This was in Birmingham, which you mightn't think an ideal place for birdwatching, but we lived a half-hour's walk from a nature reserve, and I'd go down there most weekends or after school, with my father or, increasingly, on my own. I poured into this hobby the same vast, nerdish endowment that I'd previously invested in the cars and planes, and by my early teens I could identify most British birds by sight and sound, my knowledge growing as we came across different species on trips to the countryside and coast. It probably peaked around the age of eighteen, but the interest never disappeared altogether, though my opportunities to indulge it lessened as time went on. Then suddenly, this spring, being out of work and with a bit of time to look and listen, and perhaps as a side-effect of the great bewilderment and shift of priorities that seemed to be affecting the whole country, I felt my curiosity about birds reawaken.

Although birds are everywhere, becoming better acquainted with them isn't always simple. Many are very small, of course, and though some perch openly, others are shy and reluctant to show themselves. Then there's the fact that some of them look quite similar to each other. Telling them apart is a matter of experience, and even then it's easy to make mistakes. Matching their appearance to their songs and calls makes the business of identification much easier, and is a source of satisfaction in itself, but this isn't always straightforward either, since some birds are more often heard than seen. A skylark is almost invisible at 300 feet, though its breathless bubbling song may be heard a mile away, and a swift, racing through the air above our cities, might register to the senses as little more than a bladed shriek.

But as lockdown continued into April, and even as the grip of the pandemic tightened and the national mood darkened further, it became clear from personal conversations and also from an increasing flow of articles in newspapers and on the internet that thousands of people across the British Isles and beyond were becoming enchanted enough by birdsong to want to learn more about it. Beautiful and lilting, or monotone and irritating, the sounds of thrushes, tits and finches bent in through open windows on draughts of mild air, and we asked each other about them, recorded them on our phones or simply shushed family members and called them over to listen. And even when the windows were shut against lawnmowers, next door's children or transgressive bluebottles, brick and plaster were porous to the woozy fluting of blackbirds from fences and chimneys, and to the deep purring of wood pigeons from aerials that twanged and juddered when the big birds flew off.

So the Covid-19 pandemic struck the northern hemisphere at just that moment in the natural calendar when birdsong resumes in full force after the quiet and solitary winter months. Millions of people stuck at home, the young and the old, were not just hearing but actively listening, perhaps for the first time, to the songs of birds. And many of these people discovered, as they listened to these most ancient songs, perhaps unchanged from the Stone Age, that our minds too are porous – to emotion and influence, to delight and déjà-vu, to memories of our childhood, perhaps to the childhood of our species itself. And even beyond the pleasure to be had in some of the singing, there was and remains a strange comfort in that. What the birds are 'saying' science can conjecture, but what the birdsong means to us, what we feel in our blood, is that we're not alone, that it's okay to be shunted on to one of time's branch-lines for a while, and that it's okay not to talk; it's fine to just stop and listen.

Even as the national economy – that great consumption and distraction machine – slowed almost to a halt, the breakdown of our everyday routines also invited new awareness of the circular, seasonal time that never ceases to follow its own patterns, even when we are oblivious of them, and even while the patterns themselves are readjusting to an increasingly deranged climate. In this strangest spring I repeatedly asked myself a question that I also heard from others: how would this time be looked back on? As a year of fear and trouble, when we couldn't visit our loved ones, when death came suddenly, when people went about in masks and avoided friends in the street? We're too close to it now to tell which stories and emotions will survive. But it also seemed possible, even in the grimmest days, that this spring might be remembered differently – as the time when we first heard the birds and, hearing them, began to recover an appreciation of something universally necessary but which we had somehow mislaid in our heedlessness and haste.

I leave the house for a walk and make for the nearest park, mindful that only an hour's exercise is permitted. I don't take the main road but go out the back way, into the alley where warm air and a smell of pollen and foxes rise from the flowers drawn up by the sun from the rubble either side of the humped strip of tar. There's valerian, aubretia and forget-me-nots, plus half a dozen others I can't name. One length of wall is starred almost all over by tiny flowers of the most tentative, hopeful purple. Like many people during the pandemic, I've been discovering new, unhabitual paths, both material and mental. Now that our routines have been altered, some of us flail for occupation, while for others the unreflective flood of our daily lives down narrow channels of work and duty has slowed and spread, not unpleasantly, into deltas of idleness, stillness and thought. The formula is simple and inevitable: when we attend to different things, we begin to think differently. There is surely an opportunity here that something good could come out of this time of uncertainty and suffering.

We are sometimes warned – and it's easy to believe – that our thoughts and reactions have never been so mass-produced and predictable, arising less and less from direct experience of the world, and more and more from a designedly limited but endlessly resupplied second-hand stock, pre-fabricated in the boardrooms of moguls, on chatroom message boards, in troll factories. Our attitudes are 'managed', and it becomes harder to think beyond the patterns we're encouraged to fall into. But we were not always like this. As a child I played with my sister and our friends in the alleys behind our cul-de-sac, and when I was alone I lost whole afternoons in by-ways and backwaters of the imagination. I thought I was a Ford Capri, a cheetah, a kestrel, a MIG-29. I made inadvisable medicines out of plant bits. I drew endless maps of made-up islands, and on dreary Sunday afternoons I daydreamed and chewed my roast in syncopation with the lethargic chirps of our garden sparrows. Now, in lock-down, on solitary walks, we can again explore forgotten ways and attend to the mind's weeds and margins, perhaps to find them rich with surprise and memories.

As I walk, I pay attention and receive awareness in return. The trees are renewed, and improbably beautiful. Many have new leaves, just unfurled, still a bit damp and hairy, of sour green, silver green and frosted mint. At the bottom of the road a copper beech shelters our local mob of vigilante jackdaws. They're not bothered by me, but tilt their shoehorn heads and make a soft cacophony of caws and sneezes when a raven passes high above their coral-coloured rafters. A translucent caterpillar abseils from the lower storeys of one of the lopped trees on the verge, and I turn my head involuntarily at a passing bee's doppler.

When I reach the park it's full of song, and I close my eyes the better to listen. Over the last hundred years or so, researchers have started to investigate what these sounds actually mean to the birds, and in the case of some species, careful observation has begun to lead away from guesswork towards cautious knowledge. Some calls are quite obviously intended to warn of predators, while others relate to rituals of courtship and display. Recognizing the calls and songs of even a few species of birds can delightfully enrich one's understanding of the world by revealing an almost forgotten aspect of the grammar of reality, or like one of those picture puzzles where, if you unfocus your eyes a little, a form swims up from what had seemed a random pattern. Once it has become familiar, the new awareness can sink down to the unconscious mind, so that for example when you register on some level a particular alarm note, even if you're washing up and listening to the radio at the same time, you look up instinctively to see the cat or crow that is the cause of the anxiety.

Other cues are visual: if a sparrowhawk levers itself up from the woodland to beat above the town, then the local pigeons will circle in agitation, flashing dark then silver – 'like sky sardines', as my neighbour says. This whirling movement is a defence to confuse the predator, and once we know this, a new link has been forged in our awareness of the world; next time the pigeons wheel like that, we'll involuntarily scan the clouds for the hawk. This has nothing to do with the popular image of birdwatchers 'ticking off' individual species for the satisfaction of having seen them; the pleasure here is of the deeper kind promoted by the awareness that there is meaning everywhere, and that we're part of it too.

But neither is this awareness only a matter of objective knowledge, because bird songs and calls can also strike something profound in the individual human listener. Each call not only has some functional significance for the bird itself, and for the other birds that understand it, but also creates an emotional atmosphere that can be felt by people. The calls and responses range across various bandwidths, and some speak to the soul more readily than others. Even in bright June sunshine a robin's sombre phrase can bring on a reflective mood, and who has not sometimes felt foolishly cheered by the daft laughter of park ducks? Some bird calls seem to have the power to short-circuit time and take you straight back to childhood, and in doing so abolish duration and remove you as far from the tyranny of clocks as might be possible in a culture that thinks of itself as having abolished myths. All this is to say that, without any greater exertion than sitting down and listening to birds, you may discover climates of the soul that you had forgotten existed, or that had been drowned out in the rush and clamour of everyday life. If it's preferable not to speak of the soul then let's call it something else – some novel electro-chemical stimulation. Either way, it's plainly mystical and profoundly ordinary.

Blackbird

2. Month of Blackbirds

Above all the other birdsongs of March, the blackbird's rises unmistakeable – strident and clear. In this sombre spring, when the usual noise of people and cars is absent, the song, transmitted from aerials, trellises and lamp-posts, is loud and life-affirming, perhaps outdone in volume by the tart trilling of the wren, but more compelling in its variety and the emotion it seems to contain.

Blackbird song is often described in bird books as 'mellow', but this word doesn't quite tally with the barrage of tetchy clucks, sinister whistles, apoplectic rattles and grating see-saw warbles that these birds produce in March. It's just as well, then, that the birds themselves are so conspicuous, and so distinctive in appearance and habits, as to preclude any real risk of misidentification. That said, the bird's modern English name – they were formerly known as ousels, or woozels – is itself misleading, because only male blackbirds are black. An adult male in his spring pomp is coal-black with a marigold bill and eye-ring. The black generally appears matt, and lacks the metallic tints of magpie feathers, but in certain lights it may glance a glossy scintillation, like anthracite when angled, or the ripple of sheen that sometimes flashes across a cheap black tracksuit. Outside the breeding season and among immature males the black loses its lustre altogether, and the beak colour dulls to earwax. Female blackbirds are leaf-litter brown with a generally paler, lemony bill, and speckles may show through on the throat and belly – a hint that blackbirds belong to the family of thrushes.

The family resemblance between blackbirds and our other resident thrushes also shows in their songs, since in full flow all share a quality of melodiousness that separates them from other types except perhaps some of the warblers, whose outbursts, however, tend to be quieter and more formulaic. But only a little familiarity is required to distinguish blackbird song from that of its cousins. Mistle-thrush song has an epic and declamatory quality which somehow conveys the sense that the bird has its mind on higher things. Song-thrush tunes are superficially more similar to those of the blackbird, but they're wilder and more urgent, lacking the homeliness of the latter, and without the blackbird's secondary repertoire of irate rattles and splutters.

The anything-but-mellow noises of spring blackbirds seem a good match for their temperament, which spans an emotional range from anxious to incandescent and manifests itself in a range of distinctive postures and habits. At this time of year, blackbirds never simply fly: instead, like reluctantly retired officers, they're always 'on manoeuvres', and it's easy to see from their constant agitation that for them every flower bed is a bunker, every shed a redoubt and every hedge-bottom a potential place of ambush. They have a curious gait, bounding over grass and soil with a sort of repressed frolic, then pausing, pivoting, cocking the head to listen and stab, and leaping off again. These springy bounds of blackbirds are often executed in a bending, almost horizontal posture, somewhat in the manner of a ski-jumper about to launch, and this quirk can be a useful way of distinguishing female blackbirds, at a distance, from the other common thrushes, whose stance is almost always more erect. Among twigs and other impedimenta, bounding is impractical and is replaced by a low, furtive scuttle, hardly resembling the movement of a bird at all. In fact, there's something reptilian about blackbirds, which alongside their arch-enemy the magpie perhaps betray their inner dinosaur more than any other British bird.

Near our house there's a square of old woodland, hemmed in by houses, but too wet itself to be 'developed'. Or rather, it's not a square but a cube, since its most venerable ash, oak and sycamore trees are almost as tall as the wood is wide. To be more specific still, at this time of the year the wood is a garlic bouillon cube, pungent enough to flavour the air far beyond its physical borders of blackthorn and ripped wire fencing. It really is a fine, a *majestic* stink! Entering the wood from the road, your eyes follow the path up a steep, shaded slope that is a sprouting rug of garlic leaves and the first few delicate white starbursts of its flowers. Only the narrow path of damp, packed soil is free of garlic, but step aside to let someone pass and your soles immediately squeak on young green leaves, and the weight of your crushing releases an especially glorious cloud of scent. Other paths reveal where badgers have trundled down from their sett to grub in the gardens that back onto the wood. Badgers being creatures of habit, these channels will darken and become better defined as more leaves are flattened on each nocturnal round; they also leave wisps of grizzled hair in the wire twists of the fence, and little pits in the soil where they've excavated tubers, for badgers are great lovers of garlic.

This transitional zone between suburb and wood is especially attractive to blackbirds for its richness of shelter and food, and so it's the site of much competition at this most crucial season. Because we generally attend to them so little, it can seem to us that birds' lives never really change, and that they do pretty much the same thing all year round. We may notice, after a bit of a lag, when they've stopped singing or disappeared for a while, or for good, but because our own occasions for looking out of the window are usually so few and so hurried, we may assume, for example, that blackbirds spend the whole year tweezering earthworms from the lawn or heckling the cat. But to record all the events of a blackbird's year would require a hefty almanac. Nature is constantly in motion, and infinite tiny changes are at work in every momentary illusion of repose.

The disconcerted behaviour of spring blackbirds, so evident in their voice and manners, is due to their desperation to acquire a territory of their own within which they can securely feed and breed. Back in February, older male blackbirds had begun to redefine the territories that they held the previous year. Competition comes in the form of first-year males, new cocks on the block who need to find a place of their own, and a further complication is that some females, too, may start to defend territories at this time, although these will soon be subsumed into a territory jointly held with a male. The sum of all this is an awful lot of jostling, rather as happens in that children's game in which everyone must jump inside a hula-hoop at a word of command, with a hoop being removed in every round. Although it's impossible to know for certain the qualities that blackbirds look for in a territory, prime land will generally include bountiful and varied sources of food, cover in which to shelter from predators, and perhaps a few promontories from which to sing. Urban and suburban blackbirds tend to eat a lot of earthworms and other invertebrates, so a section of lawn or flowerbed is a must. Around the garlic wood are many such gardens, and the wood itself hosts all of those insects palatable to blackbirds that are not deterred by the reeky miasma of the sprouting bulbs.

Although blackbird territories, like some of our own political boundaries, may not be marked by any physical border, they're nonetheless revealed during early spring by the behaviour of the birds themselves. Once defined in the mind of one blackbird, the perimeter of its territory is patrolled and defended against trespassers. This can get very noisy, and March is probably the best time in the year to witness almost the whole range of blackbird behaviour. Cock blackbirds face off across an invisible border, scuttling back and forth as if separated by a transparent screen. Blackbirds seem highly strung at the best of times, but during border disputes they can really let themselves go. If you linger near two agitated males, there's every chance that you'll witness all the wing-flicking, head-dipping and tail-fanning that signals a livid blackbird, accompanied by aggressive cackling and sometimes a repressed, minatory whistle that sounds forbidding even to the human ear.

This whistle is so characteristic of blackbirds that it would merit fuller description, if that were only possible. But of all the blackbird's calls – and arguably those of any British bird – this sound is the most difficult to represent in writing. I'm calling it a whistle, but this word not only misleads in its usual connotations of cheerfulness, but also fails to do justice to a sound that seems not to come from the bill at all, but rather suggests an emanation of barely checked fury that has its source in the bird's very pith. It's less a sound than a lode of malice made briefly audible, like scalding steam seething through a crack in the earth. It's ventriloquial, and more felt than heard, as if a shadow had suddenly been thrown across the blithe chatter of the spring wood. I've heard both cock and hen birds make this sound, and always in the context of aggressive encounters with trespassers.

All this hectoring and posturing is not just for show: if two blackbirds are so well matched that neither can intimidate the other, then they'll come to blows. I saw a fight this spring, right on the edge of the wood. Starting on the ground, the blackbirds leaped and fluttered at each other, clawing with their feet and stabbing with their bills. They never rose more than a few feet off the ground, and the bouts themselves, though intense, were quite short. Such disputes exact a great deal of energy from the birds, and are something of a last resort; although blackbirds will occasionally fight to the death, in most cases one of the belligerents soon gives way. So it was with this contest: the defeated male flew off, low, into no-birds land, while the victor ascended, still clucking with excitement, into a blackthorn bush and there proceeded to strop his beak across a twig, like a duellist disdainfully wiping his sword.

It's by no means only the cock birds who are caught up in this prevailing mood of anxiety and antagonism. As the month progresses, females too become increasingly intolerant, not only of other females but also of the males, up until the time that one of them is accepted as a breeding partner. Then the female applies the vigour with which she had previously defended her own patch to the mutual defence of the matrimonial estate, making constant sorties along the perimeter and ejecting intruders of both sexes with a threatening barrage of accelerating clucks. Although female blackbirds are capable of song, they delegate this task to the males, since for obvious reasons they're reluctant to draw attention to themselves when nesting is underway.

Even after they've apparently been fixed, though, blackbird territories are still susceptible to change should, for example, the tenant die, or a hen bird choose a nest site so near a boundary that her mate is obliged to try to annex part of his neighbour's turf. Still, for all the jockeying, by mid-March the pattern of territories is relatively stable, as it needs to be for the business of nest-making to begin. It's at this time too that blackbird song undergoes a number of changes, as the calls instrumental for wooing and feuding lose prominence and a more expansive set-list is unveiled.

While it's obviously important, to avoid misunderstandings, that each bird species possesses its own distinctive calls and songs, in the blackbird and some of the other more sophisticated songbirds the variety of sounds is quite astonishing. From close listening, as March gave way to April, I was able to categorize blackbird songs (as opposed to the simpler and more standard calls) into four or five basic types – but then the more I listened the more I became aware of such internal variation in these songs as to almost make a nonsense of trying to fix them in this way. Indeed, it gradually dawned on me that at least some of the birds I regularly heard from the house, in the park and at the garlic wood had their own individual quirks of speech, just as people do. It was certainly apparent that the *types* of song tended to change throughout the day – and here it must be added that the 'day' of a blackbird in spring can be long indeed. Even while the hours of daylight are themselves lengthening, it remains the habit of many blackbirds to begin singing before dawn and to continue until after dusk.

I often lingered in the wood at twilight, listening to the birds and hoping to see the badgers. I suppose it's just the ordinary effect of slowly spreading shadows, but woodland dusks have an uncanny way of seeming to begin from below: instead of withdrawing first from the sky, the light seems to be absorbed by all the matter of the wood, as though every trunk and leaf drank the light down into itself. So I stood there quietly, watching this slow encasement in shadow and listening as, one by one, the bird species ceased to call. The wood pigeons were the first to break off, followed by the nuthatches, the woodpeckers and the dunnocks. By seven o'clock the only songs left were the hesitant phrases of the robin, the sharp rapid spool of the wren, and three or four blackbirds still going strong. It was only after another hour that the last songbird fell silent and an owl gave out a quavering call from somewhere over the rise. As for the badgers, my human smell must have cut through the thick garlic odour, and I waited another twenty minutes in vain, until it was almost too dark to see.

Late to bed, blackbirds are also among the earliest risers, and it would be small wonder if their crankiness weren't the outcome of sleep deprivation. In fact, blackbirds are among many songbird species able to parse their brains' electrics in such a way that one hemisphere sleeps while the other remains on standby, lest its repose be disturbed by an unfamiliar rival or, more fatally, a hunting owl. There seems in any case some advantage to be gained by singing for so long, or else the birds would surely divert their energies into activities less likely to attract cats and birds of prey.

Given their devotion to their calling, and their robes of Benedictine black (the males) or Carmelite brown (the females), I've taken to measuring blackbirds' days according to canonical hours rather than the vulgar oversimplification of morning, afternoon and night. Thus the day begins at lauds with alarmed rattles and the odd rusty whistle, and develops in the later portions of the morning to reach its height between prime and terse (roughly 6–9 a.m.), when almost every bird chimes in with a set of more or less uniform lyrical phrases. There tends to be a lull around noon, and then the later peak comes between nones and vespers (3–6 p.m.), before the song proper peters out to be replaced by a neurotic compline of panicked clucks and the persistent metallic *pink-pink* call which apparently signals fear of the coming dark – but is explained less romantically by scientists as a token act of belligerence directed at other males who may converge to share a roost.

The overall pattern is that blackbird songs gain in extravagance as the day progresses, but even the more standard phrases of the morning are subject to sudden freaks, especially in the closing notes. One bird in our local park, singing every day from a holly tree by the tennis courts, regularly appends an indelicate whistle to the usual pattern of notes, like a schoolchild livening up a sterile writing exercise with a ribald marginal sketch. In general, the morning calls seem a bit creakier, as though the vocal apparatus has stiffened during the short hours of rest and recharge, but a certain grating quality can persist even into mid-afternoon, by which time the repetitive cadences of the morning have given way to something much more baroque.

Unmated and stateless blackbirds, in an attempt to overcome these deficiencies, sing more than their paired and landed rivals. But their desperation also seems to limit them to a fairly fixed litany of set phrases, and all day they rehearse the same lines that evolution has selected as those best fit for attracting mates and deterring other males. The more successful males, on the other hand, can afford to indulge themselves more in aesthetics. It appears that with blackbirds as also with us, when the need for labour is lessened then there's the possibility to explore the personality and engage in creative experiment to the limit of our individual gifts. This could account for the contrast I became aware of between the mechanical music-box phrases of the unmated males – who were still, as it were, being conducted by their genes – and the greater freedom of expression of the established birds, in whose unpredictable songs I imagined I could detect individual tones ranging from bullish to smug. Perhaps it's fanciful to attribute such qualities to other species, but reserve the charge until you've heard my friend by the tennis courts whose conceit knows no bounds and who finishes each phrase with such a flourish that I turn every time, expecting him to have unfurled a scroll of his title deeds.

While many of the gradual changes in blackbird songs and calls can be matched to changes in the breeding cycle, some relate to specific situations, and familiarity with these can be a great help in making sense of what's going on in the garden or the park, so that one's awareness spreads beyond the particular vocabulary of this or that species to reveal the web of the birds' interactions and wider ecology. In particular, learning to interpret blackbird calls can be a great aid to finding their predators; probably half the tawny owls I've ever seen were given away by a mobbing chorus of smaller birds in which blackbirds always seem to be ringleaders. The mobbing call is harsh and unpleasant even to human ears, so to the much more sensitive hearing of an owl each note must strike home horribly. The blackbirds' fear and anger alerts other birds, too, until the poor owl is surrounded by robins, tits and wrens, all bobbing and weaving at a safe distance. If the owl is particularly unfortunate then jays will join in too. A jay is a truly beautiful bird, but its vocal cords are made of the coarsest sandpaper, and no owl can withstand its invective for more than a few minutes. If you can tolerate the sound yourself, then sooner or later you'll see the weary owl launching silently from its branch and steering its great blunt body in utter silence to a more discreet perch in the thicker shadow of the wood, there to plot its vengeful supper of robin, blackbird and jay.

Although the blackbird is now arguably our most familiar garden bird, this wasn't always so. Its older name of wood thrush tells that its original home was in the forests, or at least on the forest margins where it's still common. Nowadays you get them everywhere. Where we live, blackbirds often forage on the beach, tossing aside piles of seaweed in search of sea-lice and other small beasts of the littoral just as they fling up the humus in woodland, or draw oaths from the gardener by scattering the leaves newly raked to the side of the lawn. At first it's incongruous to see blackbirds leaping and scuttling on the shore, but in

many ways this seems to be a perfect habitat for them, the combination of invertebrates and fruit-trees providing a year-round
source of food, and the dense, thorny brakes coming as close
as possible to a guarantee of safety for spring nestlings. From a
human perspective, it's a source of strange, inexpressible delight
to hear the fluting of blackbirds as one looks out at the ocean, just
as gulls crying far inland can sometimes stop you with a pang of
melancholy that seems to speak straight to the heart.

The blackbird's ubiquity is a good example of how nature and
culture, far from running parallel, are always touching, inter-
secting and influencing one another. Blackbirds couldn't find
much of a foothold in congested Victorian cities, but changes
in urban planning created much more congenial habitats for
them. The garden city movement, slum clearances and the Luft-
waffe's legacy of acres of wasteland within London and other
major cities played a part, but most of all the spread of blackbirds
seems to be associated with the coming of suburbia, which is
where I found them as a child in Birmingham in the early 1980s,
kneeling at the window with my chest pressed to the radiator and
my chin propped on the fake teak sill, staring out through the
endless Midlands rain at the blackbirds that scurried and flounced
on the front lawn.

Suburbia seems to suit blackbirds very well, and not only
for the ecological reason that the greenery provides them with
enough food and shelter to thrive, but also because their shriek-
ing outbursts from the laurel and their jumpy evening rituals
chime perfectly with the popular counter-image of suburbia as
the site of barely repressed human hysteria. But when I think
back to the blackbirds of that time – which is when my grandfather bought me my first bird book and my parents my first pair
of binoculars – what I remember most is their singing on warm
April evenings. It seemed that almost every available vantage,

each fence, chimney and tree-top, was host to a cock blackbird in full flow. I lay in bed listening to their voices – some confident and clear, others rollicking and wonky – and I remember being awed by the cumulative effect of the four or five birds within earshot, which seemed to me both wonderful and uncontrollable, as though April itself were flowing in the song, as though the outpouring would never cease.

This aggregate enchantment of blackbird song is perfectly caught by Edward Thomas in his poem 'Adlestrop', which he wrote in 1915, not long before his death in the First World War. The poem describes the experience of his train stopping seemingly without reason near a country platform, from the window of which he notices details that he would otherwise have been hurried past. Into that stillness breaks the song of a blackbird, and then

> Close by, and round him, mistier,
> Farther and farther, all the birds
> Of Oxfordshire and Gloucestershire.

Though everything we know about the comatose state of our countryside should warn us against reading these lines cosily, it remains marvellous that the network of blackbird song still stretches, as might be imagined, from shore to shore in an unbroken, overlapping relay. Blackbird song is loud and far-carrying, and if the sound-waves were visible then I imagine they would resemble concentric rings of vibrations centred on each singing bird, a compass of circles engraved on the mild spring air by each golden bill. Given that birdsong is primarily used for communication, each bird needs to be heard by its neighbours, and therefore these ripples of sound must blend and intermingle until the combined effect is of a great tremulous web of sound thrown across the whole archipelago, from Scilly to Unst. And at any place beneath this web what great pleasure it is to lie on the grass somewhere near where three blackbird territories meet and listen as their rivalries provoke them to ever more extraordinary vocal feats, so that the same bird you heard this morning mass-producing more or less identical short phrases now sounds more like an avant-garde flautist playing through a faulty intercom.

Where humans and birds have co-habited for centuries across greatly differing landscapes, and when the birds in question belong to a species as distinctive and numerous as the blackbird, it's no surprise that they should figure across a very broad range of cultural references. Blackbirds have been sacred to the Greeks for their song, and to the French for their flavour when roasted with juniper berries. Like other birds with black plumage, they have sometimes been associated with death or the Otherworld, as in the medieval tale 'How Culhwch won Olwen', in which Rhiannon, goddess of the dead, is accompanied by three blackbirds. But, myths being unreasonable things, blackbirds were often simultaneously thought to be messengers of love, as in the Welsh folk song 'Aderyn du':

Aderyn du a'i blufyn sidan,
A'i big aur a'i dafod arian,
A ei di dros ta'i i Gydweli,
I holi hynt yr un'rwy'n garu.

Blackbird with wings of silk
And golden beak and silver tongue
Will you go for me to Cydweli,
To ask how my love is?

The myth of Persephone may represent the richest synthesis of blackbird mythology, albeit indirectly. The link lies in the Ancient Greek belief that pomegranate seeds were fatal to blackbirds, just as they were the food by means of which Persephone's forced marriage to Hades was confirmed. It was the love of her mother, Demeter, that mitigated Persephone's fate and arranged for her annual leave of absence from the Underworld, her reappearance from below inciting the blossoming of the earth each spring, and the revived clamour of birdsong. This implied association with Persephone allows the blackbird's various and apparently incompatible connotations of love, death, mediation and hope to be combined in such a way as to make one fear for the condition of the beloved of Cydweli. Of course, even in our more literal age, blackbirds are still seen as welcome harbingers of better weather and longer, lighter days.

By the end of March the whole of the garlic wood is a trembling false floor of small white flowers that swags gently in the sea breezes which weave inland between the mossed trunks. Lick the moss and you can taste the salt, especially after one of the frequent westerly gales. I never return from the wood without pocketfuls of young leaves to bake into pastries, and several times I've seen the lady from the Vietnamese restaurant filling a basket here too; but these depredations have made no visible impact on the garlic forest that still stretches away over the slope. There's a constant traffic of blackbirds low over the flowers, and from the quiet long whistles given out by some of them I know that there are now nests nearby. In the deeper shade the wood sorrel is also in flower, and it won't be long before bluebells poke from their sheaths to add their iridescent, smoky drift to the delicate colours of the spring wood.

Later, back home, I put my head out of the skylight to listen to the cock blackbird on our terrace exchange songs with two others in the nearby park. Now that territories are beginning to be established, the songs are noticeably relaxing, and some of the variations are delightful. The most distant of the park birds is a real virtuoso, and he seems to enjoy the contrast between mellifluous, fruity passages and clownish off-key notes. His exertions inspire his neighbours to experiment with wilder compositions of their own. I call the children to listen, and for a few moments we're all quiet, our heads protruding from the roof above the quiet streets, attentive to the song that pours out of three minuscule throats. The memory returns of my own childhood in those far-off days when I would lie beneath a colourful quilt on warm spring evenings and listen to all the blackbirds of the area singing down the dusk. Those individual birds may be long gone, but the song remains, and though its full meaning may be inaccessible to us, yet nevertheless it seems in moments like these to be an essential ingredient of something we might hesitantly recognize as home.

Skylark

3. Lark Ascending

When the rest of the family go for a walk I take a break from work to look at the headlines, which are still full of warnings to stay indoors and avoid contact with other people. I escape into our stone-flagged yard and poke at some gardening, then bring a chair out and just sit for a while, listening to a wood pigeon on next door's aerial wheeze deep and husky. Somewhere further away, a collared dove, a distant relative, is also calling. His voice is much slighter and has a different tone – less a *coo* than a *cü*. Transcribing the calls of birds is a hopeless task. For one thing, since I started to pay attention during the lockdown, I've noticed that they don't seem to use consonants. We write 'caw', 'coo' and 'cuckoo', but the more I listen I can't hear any plosive *k* at all. Vowels seem better equipped to approximate some bird sounds, but here too we can't do much more than convey a vague similarity.

We shouldn't be surprised: birds have neither lips, nor teeth, nor vocal cords, and though they do have a larynx it is not for them the 'voice box' that it is for us. Instead, birds have an organ called a syrinx, named after the nymph of Greek mythology who was transformed into reeds from which the first 'pan-pipes' were made. Given that the mechanism of bird vocalization is so different from our own, there's really no chance of trapping it in an alphabet. And if we can't even get 'coo' right, what chance do we have with the more elaborate song of a thrush or a wren?

Some bird calls are more complex than others, of course, and this difference has an anatomical basis in the position of the syrinx within the birds' overall vocal equipment. All the species that we call songbirds have their syrinx between the windpipe and the tubes that lead to the lungs, and because these tubes ('bronchi') fork in a 'X' shape at the point where they branch into their respective lungs, these birds are able to produce sound with air drawn from either the left or right lung, or both at once. Within the syrinx are membranes of thin but tough tissue that vibrate when air is drawn over them, and it's these vibrations that produce birdsong. More accurately, they *are* birdsong, and I find this helpful to remember whenever I reflect on the fact that our culture, if it admits the spiritual at all, insists on opposing it to what is physical and fleshy. Poets have rhapsodized for good reason about the power of birdsong to sound neglected parts of the human soul, but just as it's sometimes well to consider the spittle in the flute or the sweat on the strings that draw music from man-made instruments, so it should enhance our appreciation of birdsong to recall that its beauty has a physical basis in the exertion of tiny lungs, muscles and membranes.

It's the ability to switch at great speed between different bronchi (and thus different sides of the syrinx) that allows for the astonishing sophistication of birdsong. Different notes, pitches and tones can be emitted from each bronchus, which allows for transitions of sound far beyond anything achievable by even the most virtuosic human musician. Of course, this applies only to the sounds made by songbirds: the more basic calls of ducks or swans, say, have their origins less in the syrinx than in the convulsion of muscles within the trachea itself. The fact that scientists have used the possession or lack of a complex syrinx to categorize all birds into one of two classes – the 'oscines', or songbirds, and the 'non-oscines' – is one indicator of the significance of birdsong to our own species.

Although it's obvious on reflection that birdsong is only as effective as bird hearing, this tends to be overlooked by ordinary listeners and specialists alike, who focus on the singer rather than the listener. This is understandable – but birds' sense of hearing should nevertheless be taken into account by anyone who desires a greater understanding of birdsong, for the complexity of bird speech and the subtle discrimination of bird hearing are interdependent.

Though they're invisible behind feathers, birds' ears are in most respects quite similar to our own, the main difference being that birds are capable of a far superior refinement when it comes to distinguishing between rapidly uttered sounds, so that what we hear as a one-note trill may be received by its intended listeners as a complex series of different notes. In other words, much more information is conveyed by birdsong than we'd credit – although to think of birdsong as a mere means of sending information would be unhelpfully reductive. The rationalizing human tendency to separate content from form and message from medium shouldn't be applied to birds, for whom the distinction almost certainly doesn't apply. Instead, birdsong is an irreducible synthesis in which *what* is sung cannot be abstracted from *how* it is sung. In this respect, birdsong is immersive and immediate in a way that, in human terms, is much more suitably compared to an artwork than an email.

When thinking about birdsong, then, we should take care to avoid falling into the trap of assuming that birds hear themselves in the same way that we hear them. Just as what sounds to us like a single slurred note might be richly various for the bird, if we hear certain songs and calls as repetitive or monotonous, it may well be that we simply lack the ability to discern the small changes that speak eloquently to other birds. Not that our sensory deficiency should deter us from listening more carefully, of course, for there's ample delight in the birdsong available to our ordinary attention. There's a robin at the end of our street, a streetlamp crooner who only really gets going at dusk. When I first began listening to him I could only distinguish three or four phrases that seemed to be repeated, often in the same sequence, over and over again. Now, two weeks later, I've lost count of the subtle variations that leaven his song, and if my imperfect hearing can do as much then it seems certain that the ear of another robin could discern far greater intricacies.

The question of whether birdsong is properly comparable to music, being appealingly beyond hope of proof, has sustained a vigorous, centuries-old debate that shows no signs of petering out. The most secure argument of those in favour is that birds employ the basic elements of pitch, rhythm, intervals and variation that constitute the grammar of music; meanwhile, there's also plenty of laboratory evidence to support anecdotal observations that some species can improvise as well as any jazz musician.

A common counterargument is that the comparison of birdsong to music is wishful thinking, and far too hard on the birds. Birds aren't confined by the scales of notes that are the basis for our music, and have available to them a far greater variety of sounds. Some birds, such as the song thrush, can even overlay a set of notes produced from one side of the syrinx on to a different, lower-pitched set emitted by the other side. It is as though you or I sang melody and harmony at once, drawing from a nearly limitless number of notes, at a rate of up to forty notes per second. Given this, to bracket ourselves alongside thrushes as musicians seems like a typically human piece of hubris.

Nevertheless, the musical qualities of birdsong have often been noted and appreciated by human maestros. The great Slovakian composer Leoš Janáček kept a pet goldfinch for inspiration, and

Mozart himself marvelled at the prodigious performances of his tame starling. Beginning in the 1940s the Frenchman Olivier Messiaen, a keen ornithologist, heard enough similarity between human music and birdsong to incorporate the latter into his compositions. His synthesis of bird and human music reached its climax in his 1958 piano piece *Catalogue d'oiseaux*, which features snippets of song from more than seventy species of bird.

But if this is birdsong *in* music, instead of birdsong *as* music, might it be better instead to think of birds' sounds as language? The best, though least satisfying, answer to this question is: 'it depends what you mean by language'. Nobody who has watched birds for any time at all could reasonably doubt that they're communicating with each other. A few minutes ago, while I was wondering about this, a large seagull landed on next door's shed and began making the nervous, rapid *ak-ak-ak* that meant it had spotted the cat glowering on the opposite wall. No doubt the sound reflected the bird's own anxiety, but it was also understood by all the other gulls around. More than that, though the finer nuances of gull-speech may be reserved for other gulls, the basic message of alarm is certainly recognized by other species too. Sometimes a buzzard drifts over the street, and when it's spotted the local gulls start up a summoning honk to call for reinforcements, quite horrible to the ear. A starling or jackdaw caught by a predator or trapped by a human being will make distress calls that seem to have the same function.

These examples might be termed calls rather than songs proper, and this distinction has often been used to separate those bird sounds we think of as linguistic from those we may be more tempted to praise as musical. But there's more than might be imagined to even the shortest and simplest-seeming calls. Among many different species of songbirds, for example, the alarm calls for an overflying hawk are almost identical: a thin, compressed whistle, delivered with the bill barely open. To ask why is to invite a partial and over-simple answer unless the birds' whole environment is taken into account. In the case of these hawk alarms, it's been noticed that they sound the same because they're expressed at an almost exactly similar pitch. Experiments have shown that this particular sliver of pitch, although loud and penetrating, is especially absorbent by branches and foliage – a discovery suggesting that the combination of loud volume and ventriloquial effect is designed to alert as many birds as possible without causing excessive risk to the sentinel. The hawk can't get a fix on the source of the sound, and to the human ear, too, the sound-print is smudged and hard to trace.

This alarm call, then, has a social effect without necessarily having a social intention, but other bird calls can only be understood in the context of their social lives. Among these, the most frequently heard are those that birdwatchers and scientists usually label contact calls. In a corner of our nearest park there's a sombre stand of pines where it's often possible to see kinglets, birds that love conifers more than any other kind of tree. The high-pitched piping of these tiny creatures – also called golden-crested kinglets, or simply goldcrests – is always disproportionately loud in relation to their size, but yesterday when I approached the pines from the path their calls were all the clearer for the amplifying effect of the ornamental pool that divided us. I awkwardly crept round the muddy edge, but then slowed and approached the trees carefully, although these birds aren't usually shy of people: it may be that we're too large to be made sense of, or simply too clumsy to seem a threat. I stood still to watch them fizzing about one russet, fissured trunk, tilting their heads to squint for spiders' eggs in the darkness between needles. But the longer I watched, the more I came to doubt that I could see them move at all. It's always like this with these birds: they're so quick that they're simply there one second, and somewhere else the next, as when an image blurs above the pages of a flick book, and the brain is forced to fill in the gaps. They came nearer, and for a minute or two they were all around, haloing me with their thin silver calls that seem almost as much of light as of sound. And how else could such tiny animals avoid losing one another in this vastness, but with voices that flash out into the dark? The birds are scarcely larger than a ping-pong ball, and weigh perhaps only twice as much. I think of the goldcrests' sesame-seed brains, flickering with the electricity needed to pilot their restless bodies through this immense darkness. For a single pine tree must be like a vast, unlit city to a kinglet. And if this is so, then wouldn't a thicket be a world, and a forest of pines a dark nebula through which they are driven by the search for the food, but also by the awareness of how vulnerable they would be if they lost contact with their fellows?

It isn't only very small birds that use such contact calls to keep in touch. While their heads are down grubbing, playing-field jackdaws, intent on insects, will nevertheless maintain conversation with others in their party by a constant patter of squeaks and pops and genial caws, while on winter nights migrating thrushes flying exhausted above the city encourage each other with brief pulses of sound, making momentary meaning in the interstices between stars.

Despite the increasing acceptance that birds aren't just the puppets of their genes but can actually make choices about how and what to sing, deterministic explanations for birdsong are remarkably persistent. Part of the great appeal of such beliefs is that they really can be made to explain almost everything; there is virtually no action a bird can perform, however seemingly unnecessary, that would forbid an evolutionary 'solution'. For example, the theory maintains that birds are driven to sing in order to attract mates so that they can continue their line. Someone objects that robins continue to sing long after the breeding season has finished, to which the reply comes that robins are unusually aggressive; they are compelled to maintain their territories throughout the year, still to the end of maximizing their breeding success.

As research into birdsong continues, however, aspects of behaviour are coming to light that seem to fit into a strict evolutionary schema only tortuously, if at all. While robins' late singing and general truculence might be explained by their territoriality, this neither accounts for why other bird species *don't* do it nor satisfies in the case of blackbirds, for example, which also sing late into the autumn, despite the dissolution of their territories.

To deal with this problem, some biologists have asserted that late-singing blackbirds are practising for next spring, and thus trying to steal a march on their rivals – but this argument looked questionable even before closer listening revealed that the autumn singers very often extend their compositions far beyond their breeding season repertoires. Yes, certain blackbird songs can plausibly be linked to courtship and defence of the realm, but mostly these aren't the songs that are 'practised' in the autumn. Indeed, what blackbirds sing beyond July is often so exuberantly different that it's hard to avoid the conclusion that they do it because they enjoy it. Of course, blackbirds' enjoyment of their own songs doesn't disqualify an ultimate evolutionary motive, but neither does it obviously require it. The least that should be conceded, I would say, is that blackbirds are emotional animals that don't simply react to stimuli but also make choices and develop preferences with regard to what they sing, and no doubt much else besides.

Blackbirds aren't the only species that seem to relax and expand their repertoire at a time of the year when the sex drive is at its weakest; the same phenomenon has been noted among some warblers and larks. If these summer embellishments really are as unnecessary to breeding success as they seem, then it must follow that the birds derive personal satisfaction from freestyling in this way. We could take the hypothesis even further, and turn scientific convention on its head, if we were to hazard that these species sing so freely in late summer precisely *because* they're now free from the necessity of repeating the set phrases that make up their recognized courtship and territorial songs. When one has finished writing an essay, or a work email, one is free to play with words just for the fun of it. Might this also be the case with birds?

Some of the most joyful phrases in the English language are the collective names for groups of birds, many of which are quite obviously emotional in their associations. Consider a 'piteousness' of doves and an 'exaltation' of larks. To the objection that these are projections on to the birds of *human* emotion there is the equally obvious retort that the word would not have stuck if it didn't correspond persuasively to our sense of what really is going on inside the bird.

Probably only the nightingale's song has been more celebrated in British culture than that of the skylark, and even then it's a close-run thing – in a 2011 survey for a classical radio station, Vaughan Williams' *Lark Ascending* was voted the most popular piece of music. Like those of almost every other farmland bird, lark populations have collapsed in this era of agribusiness, but a half-hour's bike ride from our house there are still good numbers of them in a strip of rough pasture behind some coastal sand dunes.

Of all the landscapes on these islands I love wet woods the best, but second to them I love dunes. And while I love woods for their intricate, infinite complications of perspective, terrain and shadow, with dunes it's their simplicity of form and colour that makes it so pleasurable to walk through them or even just to lie and look up at them from the beach. On hot summer days the light reflected back from the sea mingles with the glare thrown back from the silica of billions of sand-grains. Above this blond light there's the pagan blue of the sky, and that's all the colour there is. Crowning the dunes, there's a gently swaying sea of coarse grass, like tousled hair on the temples of a hero, and that's all the *movement* there is. Lying here listening, for a long time I hear only the rhythmic exhalation of waves, but then suddenly I catch a birdsong, thin and guttering, as if it too were being blown by the wind, as when a trickle of water is pushed and plaited into new shapes by a sidelong breeze. The sound rises, seeming to twine around itself as it ascends, braiding upwards apparently without any material support, like the Indian rope trick accomplished in music. Because there's no doubt about it: this is clearly music of a strange and magical kind. The skylark itself is invisible in the white blaze between dune-light and cloud-light, but the song trembles in frail strength, consistent beyond the oscillations of its pitch. The human ear follows where the eye fails, and the heart with it, up the same giddy rope of song to where the bird itself is felt to be. An exaltation of larks – that feels about right.

As it happens, Charles Darwin himself suggested that birds are moved by emotions and may sing from 'mere happiness'. As often happens with founder-figures and their acolytes, many of Darwin's later followers were more doctrinaire than their guru, and some were unwilling to grant that birds have feelings, even when it was pointed out that birds' emotions, far from derailing the science, might be the necessary motivators of behaviour whose ultimate justification could indeed be down to successful reproduction. If birds didn't enjoy singing, then why would they do it at all? On the other hand, given that in almost all British songbird species female birds sing less and more rarely than the males, are we to assume that they are 'unhappy', or that they find their happiness in other activities such as nest-building? Whichever way you look at it, attempts to second-guess birds' motivations are always likely to intersect with contentious subjects in human culture.

If birds' sounds *were* programmed and unvarying, then this would leave no room for experiment and play, but it has long been understood that birds can also learn songs from experience. This was better known in earlier times than today: from antiquity, the owners of caged birds knew it was possible to teach them simple melodies and phrases of song, and in the Middle Ages tiny wind instruments, called bird flageolets, were designed especially for this purpose. In the wild, too, while most birds seem born with their species' integral repertoire of functional calls, their longer songs are at least partly learned and need to be practised. Young birds learn snatches and sequences of song from their parents, and perhaps also from other birds of the same species. Sounds made by *other* species may also be briefly incorporated at this time before losing favour later on, suggesting that in young birds as with human infants a loose and experimental phase of vocal development precedes the establishment of a set adult 'vocabulary'.

Many songbird species also practise so-called 'subsong', a relatively unstructured babble of sounds too quiet to be intended for broadcast. I've often been intrigued by an unfamiliar song coming from the middle of a thicket, and stood there in impatient excitement until the singer was revealed as a robin, blackbird or finch instead of the incredible rarity that my imagination had worked itself up on. Even crows are capable of the most delightful prattling monologues, utterly different from their customary black retching. The question then arises why crows and other birds with a similar capability don't make use of the full programme of sounds they have at their disposal. There seems to be some unused capacity in these cases that's as much of a puzzle from the point of view of evolutionary science as the over-elaboration of other songs. Once more, any honest assessment of human knowledge of birdsong would have to conclude that we still have a very long way to go before we can make sense of everything, and in the meantime we might do well to loosen our assumptions about birds' capabilities and motivations.

Whether through calls and song or through non-vocal means such as the belligerent wing-clapping of wood pigeons or the anxious tail-flicking of moorhens, birds communicate what is required for their survival, and to an extent their different means of communication help define the sometimes flimsy borders between species. But beyond the differentiation of species there is also in the improvisations and preferences of individual birds something that we should perhaps not be quite so hesitant to term personality.

Wren

4. The Wren and the Rostrum

After more than two weeks of calm, warm weather, in early April our archipelago is once again raked by wind. I was woken in the small hours by a percussion of shed doors rattling, plant pots rolling and wheelie-bin lids being raised and slammed. Now little eddies of dust and fallen fuchsia petals are racing round our yard, and next door two pegged-out wetsuits perform a twitching *danse macabre* on the washing line. Looking out at the neighbour's garden, I watch the wind chase the sunlight so boisterously that the flowerbeds seem splintered into shards of peat-black shadow and whipped pools of soft golden light. Further down the terrace, one yard has a vegetable patch, and the visible parts of the plants are all reacting madly to the wind. Leek-tops flail, beanstalks buckle, and the gleaming green-black kale is reeling in all directions at once. The overall effect is something like an orchestra in crescendo, with each plant fully animated after its own fashion, and the gale itself as the inspired, tempestuous conductor. Above the gardens and the terrace of houses, gulls are tossed about in the last remaining column of blue air; then they glow white as aspirin against the dark cloud-mass over the ocean, before showing dark again where they are briefly silhouetted against the dazzled ramparts of white cumulus racing to the east.

The wind makes things difficult for winged creatures, and most are lying low. Two bees career past on drunken vectors, and butterflies are nowhere to be seen, though I imagine them clinging to the bottom of leaves like desperate yachtsmen tacking into a storm. Yet when I open the window to refresh the house I immediately hear, above all the blowing and rattling of the gale, a silver wittering from the dunnock that always seems to sing from the same topmost twig of the thorny shrub opposite my bedroom window. Dunnocks are unobtrusive little birds, vaguely sparrow-like, but slimmer in every department, from bulk to beak to song.

Although it's most unscientific, I hold to a loose hypothesis that you can tell what a bird eats by the sound it makes. I daren't submit this idea to much scrutiny, but isn't it true that blackbird song is full and fruity like juicy worms or spoiling apples, while swallows and martins have thin, insectivorous voices? And that the very way we represent the calls of sparrows with 'chirp' is suggestive of a seed, with its fat full vowel sound tapering at either end to those sharp little consonants? At any rate, if there's something of husk and chaff about a sparrow's rounded chirping, then this is at least one reliable way of distinguishing it from the mostly insect-eating dunnock, whose song is by comparison insubstantial yet also strangely insistent, like an advertising jingle for a diet of lice and spiders.

I stop at the window to watch admiringly as the bird, although constantly buffeted and obliged to change its footing, clings tight to its favoured twig and continues to scatter its small change to the wind. Why does it persist with this chilly perch when it could just as well descend into the tranquillity of the bush? Is it so urgent to sing, and who's listening anyway in this roaring air?

The question of why birds sing is closely related to those of when they sing and where they sing from. Many of our most familiar species have their favourite perches for singing. Ornithologists call these special places song-posts, though here 'post' should not be taken too literally, but rather in the looser sense of an appointed place. The favoured site varies from species to species, and also seemingly between individuals. Dunnocks like hedge-tops, while garden warblers prefer deep cover; starlings enjoy aerials and robins like low perches in the undergrowth. Some birds sing close to the ground, while skylarks make their perches in the upper air itself. The importance of song-posts is relatively understudied, and among humans it has probably been most fully exploited by makers of greetings cards picturing a robin on a spade handle, or a kingfisher posing with a minnow on a 'No Fishing' sign.

Leaving aside questionable human tastes, song-posts are clearly very important to birds, and many sing shorter and less varied songs if their favourite place is inaccessible. Some birds' predilections are so marked that they've become legendary, such as the great tombstone in the churchyard of the Welsh village of Nanhyfer (Nevern) in Pembrokeshire, associated with the sixth-century saint Brynach, from where the first cuckoo of spring was said to call each April.

There are no cuckoos around here, but, still thinking of Brynach, I put on a coat and walk the short distance to the park, to listen out for other birds that may be braving the gale.

Can the blind discriminate tree species by the sound they make in the wind? Even my inexpert ear begins to notice, after a little attention, that the eucalyptus is sibilant in a way that's different from the deeper dry torrent of the beeches, or the papery rustle made by the unkempt dreadlocks of the willow by the ornamental pond. The wind makes easy play with the small-leaved, loose-limbed trees, but those that hold their branches closer to the trunk or, like the horse chestnut, have leaves that are large, lobed and relatively few, make almost no sound at all in even the most powerful gusts. It's surprising, too, what different emotional tones can be suggested by the mere action of the wind on trees. The maples planted along the playground fence give an impression of distress as their loose leaves flap and dither on weak stems, but the three far poplars take on a new grandeur. With their wild swaying slowed to half speed by the effect of distance, and their flashing oscillations of pale green and silver, they somehow give a watery impression, suggesting trees less than tethered waterspouts, great oscillating columns of pale green and riffled tinplate.

With the trees temporarily inhospitable, the birds are having to adapt to the new conditions, and most have dropped down to the quieter microclimates of the understorey. The blackbird that usually sings from the top of the dead ash tree has relocated to the denser cover of the magnolia, and the goldfinches have abandoned the alder crown to seek seeds newly shaken from the weeds around the tennis court. But I'm surprised to discover how little the park birds, like my backyard dunnock, seem to be inhibited from singing by the wind. The dark interiors of laurel and rhododendron are full of fluttering, twittering life: blackbirds still grate their innuendoes, robins repine from the shrubs – but loudest of all is the wren, bobbing and irate on a beech stump, for all the world like a little tinpot dictator bellowing from a rostrum.

I often see this wren singing in precisely this place, and it leads to the thought that what for humans may seem like a more or less undifferentiated jumble of stalks and sticks may for the wren be a much more considered thing. I can only see these shrubs from the outside, which makes my perception of them literally superficial. Not only can I not see the wood for the trees, but I can't see the twigs for the bush. Moreover, since I'm not used to noticing them, I'm aware of my eyes slipping off, round and past the bushes to the gap beyond where the light is combed through the unmown grass of the playing field.

Birdwatchers, of course, can learn a lot about birds by watching them, but no doubt they can also learn a lot about watching – I mean the act or process of watching itself, and why we see what we see and miss what we miss. It's a basic premise of art criticism that our visual priorities are trained and governed by convention; the convention of perspective, for example, is so ingrained as to seem natural and unquestionable, which may well help explain why my eyes are constantly restless to escape the complicated detail of the foreground. But that's no reason to assume that this wren, innocent of the Renaissance, sees things the same way. I'm also perhaps simply too large to appreciate the bush as anything other than a bush, whereas for the wren it combines the functions of hotel, supermarket, auditorium and air-raid shelter against marauding hawks.

Places matter to birds in all kinds of ways, many of them specific to individual species and by no means only to do with song. Within their territories, some birds favour particular feeding stations. Song thrushes enjoy snails, and unlike blackbirds they know how to get at them. Picking up the snail by the cusp of its shell, the thrush proceeds to hammer the poor creature against a stone or stone-hard root until the shell breaks and the defenceless and dazed invertebrate is swallowed. This practice means that in areas where snails comprise the larger part of the thrushes' diet the birds return again and again to wherever their preferred snail-stone happens to be.

Even where there's no such obvious reason for a bird to haunt a particular site, it often happens anyway. Birds of prey such as sparrowhawks, having caught a small bird, will sometimes carry it some distance to a so-called 'plucking post' – if you find one, its purpose will be obvious from the ring of small feathers let fall around the stump, and if you're especially lucky then maybe also the odd gobbet of gore.

Other physical signs of birds' favourite places are the droppings deposited at roosts. If you've ever left your car parked under an ivy-cloaked tree when wood pigeons come in autumn for the ripe berries then you'll be familiar with this. Watching the big unwieldy birds flop and totter for the berries on twigs far too slender to support them can be entertaining, but you'd have to enjoy it very much indeed to judge it worth spending the next morning sponging purple guano off the windscreen. Unlike sparrowhawks, owls swallow their food whole, and they deal with the indigestible parts by regurgitating them in little lozenges of feather, fur and bone. Should you find an owl roost, and if you can stomach it, then you might gather up some of these owl pellets and take them home for dissection. As a boy I assembled in this way quite a collection of vole skulls and the glossy bramble-black wing-cases of beetles, to the unexpressed delight of the rest of the family.

The most obvious dilemma for birds in search of a song-post is how to maximize their songs' reach without exposing themselves to unnecessary risk, but within this broad trade-off are a great many hidden complexities that have as much to do with an individual bird's local environment as longer-term adaptations. A bird wanting to minimize its chances of being killed needs to have some sense of which predators should be most feared within a given space, since it would be counterproductive to avoid high perches, for example, in places where there are few sparrow-hawks but many cats.

Climatic and weather conditions also affect how much energy is required to sing and how far the song will be likely to travel. The birds singing into the wind today clearly deem it worth-while, but if the air were colder, or if it were raining, then the extra metabolic exertion might be forbidding. It's also worth considering why a bird is singing in the first place: while some studies have suggested that female birds might be attracted to 'risk-takers', there can be no advantage to broadcasting the exact location of the nest once mating is over. Skylarks walk a little distance from their nests before they ascend to sing, and it would be surprising if other birds didn't take similar precautions.

Keen interest breeds expertise, and birds are so invested in their singing that it makes sense for them to be highly attentive to their sonic environment. What is for us the niche interest of musicians and sound engineers is for songbirds a matter of life and death. But while sound engineers can control almost every aspect of the studio environment, birds must work with what they find. For thousands of years, birds' sonic contexts must have altered very little, but that all changed in the last few centuries, as humans began transforming the natural environment into the built environment.

At the bottom of the alley that runs behind our terrace is a pathless, polleny wilderness of native weeds and garden exotics on the run. It's impenetrable to human beings, though going by the flattened channel through the metre-high nettles and the giveaway pong it's clearly no barrier to foxes. Beyond the nettles there's a shapeless agglomeration of holly, crab apple, ivy and flowering currant, all growing against some dilapidated garages that they seem to be simultaneously supporting and wrestling to the ground. Somewhere within this thicket is Sparrow Central, the neighbourhood headquarters from which the birds quest in all directions for food. An incessant cheeping emerges from this tangled estate even when the birds are hunkered down in bad weather, but during April I noticed that if a sparrow had something particularly important to say, it would leave the vegetation and fly up to the guttering of one of the nearby houses. I soon realized that these birds were probably prospecting for nest sites, and it occurred to me that they flew up there so that their territorial boasts would be amplified by the plaster and brick of the house-fronts. So loud was some of the chirping in the trafficless street that it bounced back in an echo which, having nowhere else to go, was funnelled up the terrace. The result was that a sparrow broadcasting from the guttering of no. 2 sounded almost as loud at no. 44.

A growing amount of research demonstrates that urban birds sing differently in response to the sonic properties of their man-made environment. Partly this is to do with volume. Normally this town, like any other, is dominated by motor traffic and other human noise – so the birds need to sing more loudly to be heard. A country robin receiving a visit from a townie cousin would probably find himself having to retreat a twig's length so as not to be deafened. Some urban nightingales, too, apparently reach such a volume that they breach European Union directives on sound pollution.

But it isn't only traffic noise that obliges birds to change how they sing: pitch is also important. The rumble and drone of engines occupies the lower frequencies of the sound spectrum, effectively forcing birds to sing higher if they want their voices to carry. Perhaps one reason why wood pigeons are seldom found in city centres is that they can't hear each other; so they restrict themselves to leafier suburbs where their smoky baritones are more effective. Feral pigeons coo at a slightly higher pitch, but they also tend to be more sociable than their woodland relatives, meaning that they can communicate within the flock without the need to project their voices very far.

Pigeons are in any case fairly monotone, but species whose songs range across a variety of pitches are faced with the slightly different problem that only part of their repertoire might be properly audible, no matter how they exert themselves to sing louder. This issue is particularly acute for species such as the great tit whose females seem to have a bit of a thing for deep-voiced males. What's a male great tit to do, when his winning low notes are obscured by the revving of cars and buses? Clearly, some sort of compromise is required, and while it's likely that the birds switch to higher registers in noisier parts of town in order to be heard at all, the females' selective predilection for deeper voices may remain a decisive influence where traffic is lighter.

But as in the case of our terrace sparrows, not only traffic but also the urban landscape itself plays a part in how and where birds sing. Compare a zone of urban space with an acre of woodland. A healthy woodland is a multi-storey chaos of trunks, branches, twigs and leaves. These objects all absorb sound and, being of different shapes and densities they also baffle, cushion and deflect it to variable degrees and in more or less unpredictable directions under the effects of natural growth and the action of the weather. In contrast, the city plot is dominated by straight lines and reflective surfaces from which birdsong redounds in ways that the bird can come to know. This sonic environment may well provide another incentive for birds to sing at a higher pitch, since a higher note produces fewer echoes to impinge on the clarity of the succeeding one. All these factors might be at work in the mind of the sparrow at no. 2, and more besides. This house is on the street corner, so he is projecting his chirps in three directions at once. Not only that, but it also seems reasonable to assume that the guttering itself has an amplifying effect, for what is a gutter to a proprietorial sparrow if not a prodigious megaphone for the purpose of projecting his voice?

Years ago, when I lived in London, I would lie awake listening to a pair of robins that sang late into the night, long after the last of the other birds had turned in. The song-post of the closer robin was on the scabbed bough of a plane tree, a great brawling thing, like a docker's arm, that was illuminated throughout the night by a cone of dirty yellowish light beamed down from a corner streetlamp. Meanwhile, his interlocutor sang from a sapling outside a takeaway whose neon sign lit up the upper branches strongly enough to throw a lattice of shadows down on to the trunk. Why were these birds up so late?

Most songbirds are stimulated to sing by changes in the level of light, hence the famous 'dawn chorus'. The presence of artificial light at their song-posts might have incited these robins to sing, but it doesn't necessarily explain why they chose to carry on into the night, instead of retiring to roost. However, if we subscribe to the thesis that most birdsong is territorial in intention, then it might certainly make sense for robins, and other songbirds such as nightingales, to sing at night. The air is often calmer after dark, enabling birdsong to travel further, and in built-up areas there's less traffic too. Town robins do continue singing later than their rural counterparts, and it may simply be that unmated urban robins, or those still to fix a territory, are unable to make themselves heard during the day, whether figuratively because of the competition they face from established males or literally because their voices are lost in the traffic. These less fortunate birds may then sense that they need to continue singing after dark, but this surely comes at the cost of less time for sleeping and feeding. It may be that in the capitalist update of the nursery rhyme, cock robin is killed by the exhaustion of mandatory overtime.

In the end, as with many aspects of birdsong, after considering local specifics and attending to the birds at first hand, the best we can probably do is rank our explanations in order of plausibility and then admit that we don't yet know enough. The motivation of the neon-doused robins that sang me to sleep in London has not yet been fully explained by science, and though more knowledge would no doubt be a good thing, more whimsical explanations may not be entirely worthless. Aren't city-dwelling humans, too, overstimulated by light and sound, sometimes tempted into unnatural behaviour? And don't we also do things in these conditions that we don't fully understand, maybe even to the extent of singing lairily under a streetlight after dark outside some takeaway we'd never even glance at in the daytime?

In drawing a distinction between birds that sing at night and those that go to roost, I wouldn't want to foster the idea that the latter settle down quietly. Yes, when birds finally prepare for sleep they are generally silent, but the phase between reaching the resting place and actually going to sleep can be noisy in a way that will be familiar to any parent of young human children. I've already mentioned the clamour that blackbirds make when going to roost, but wagtails, starlings and rooks are among many other species to indulge in a similar burst of wild energy before lights-out.

As a teenager, after winter nights out in Birmingham city centre, I would wait at a bus stop directly opposite a roost of pied wagtails, in one of the very few trees in a comfortless cityscape of concrete buildings, distastefully illuminated by sickly amber lamplight crammed with cold slanting rain. There were so many birds in that tree that in mid-January it looked as if it were still in full leaf, but in my less than sober condition that may not have been enough to make me take notice if it hadn't also been for the excited noises made by the wagtails coming to roost, and the occasional outbreaks of aggravation if an incomer tried to squeeze on to an already fully occupied branch. I can't imagine what it was, because natural predators weren't to be found in that wasteland, but one night something disturbed the roost sufficiently for the whole mass of many hundred wagtails to ascend suddenly into the grainy air, like a coil of dirty smoke burning off a blown wick.

I'm brought back to the present by an especially delinquent gust that flings up a pile of last year's leaves to whirl around my boots.

The wren is taken aback, and vacates his song-stump for the basement of the wood, where he'll bustle around, mouse-like, until the weather calms and he's seized again by the urge to sing. Through the gap to the lawn I watch how the wind acts on a huge pyramidal fir, and notice for the first time how the patterns made by the racing air reveal the arrangements of branches at different levels of the tree. Around the bottom of its cone the down-swept branches overlap in voluminous bustles that rise and swell with the wind, like Marilyn's skirts above the Brooklyn vent. Further up, near where I picture the midriff of the hidden trunk, the foliage moves in roughly circular oscillations, as if it were being rubbed dry by a huge invisible towel, while at the very top the slim peak nods like the tip of a wizard's hat. And on the tip, most improbably, stands a greenfinch.

Unlike the thrushes, the pigeons and even many of the warblers, each of our common finch species sounds quite distinct. This makes identifying them pretty straightforward, and anyone wishing to recognize birds by their songs, having grasped the blackbird, robin and wren, would be well advised to next take up finches. These days, at least in cities, goldfinches are probably the most common and gregarious members of the family, and their call is an unmistakeable little geyser of bubbling bells. At the other extreme is the bullfinch, which has a diffident repertoire of unassuming creaks and croaks, as if something hidden in the woodland badly needed oiling.

Most difficult is the chaffinch – not because they're particularly timid or quiet, but because of all our birds perhaps only the blackbird and the song thrush have a greater variety of calls. After nearly forty years of listening to birds I'm still often fooled by chaffinches, and it would be hard to say that the pleasure outweighs the exasperation when ten minutes' slow stalking through ferns and nettles reveals a bird that would probably have come to my feet if only I'd thrown down a crumb in the car park. Many chaffinch calls are frankly a bit uninspiring, but they do have an endearing one that makes them sound a little mad, where they follow a querying *weep?* with a couple of decisive *pink-pink* noises, as if they were answering their own question. You imagine them nodding to themselves in affirmation, but then after a short pause they ask the same question again, and so it goes on.

The greenfinch call is much simpler, although at this distance and what with the wind I can't tell whether the bird on top of the fir tree is making it right now. But you'd know it if you heard it, because it's a single, drawn-out wheeze, passably imitated by sucking your teeth for about three seconds. Sometimes the wheezes are interspersed by a series of rising chirps, as though the bird were ratcheting itself up ready for the next long note. As well as this most characteristic sound, the greenfinch has a short call that it uses in flight, and it's in this business of flight calls that my claim about the ease of distinguishing between finches breaks down, because in motion one species can sound very much like another. Even goldfinches often replace their tinkling with a monotone *peep*, uttered at the apex of their undulating flight, as though this peeping were a necessary aid to their buoyancy.

Apart from the bullfinch, all of our common species like high perches, and if you see a small bird right at the top of a tree, it's probably some kind of finch. You don't have to walk far in any town to find goldfinches babbling on an aerial, and although my greenfinch is exceptional in his tolerance of the gale, his preference for the tip of a conifer is quite typical.

In saying 'his', by the way, I'm again aware of the danger of trimming female birds from the picture. Male finches may have their treetops, and cock wrens their rostrums, but at least in spring it's clearly just as important to determine where the hen birds are listening from, for it's they who drive the whole business of selection through their choice of mate. There might well be much greater and more complex discrimination to their listening than is required of the male in search of a song-post. An unmated female may audition many different suitors, and thus need to settle at an equidistant site where their voices overlap, but which also affords her cover and a quick egress from danger. All of the characteristics of the sonic environment – the dispersal or channelling of sound-waves, the absorbency or resonance of vegetation or brickwork, discussed earlier in consideration of the singing males – pertain at least equally to the listening females. In brief, it's a point often overlooked that birdsong is only as effective as bird listening, and this is perhaps another instance of bias by male birdwatchers and scientists towards male songbirds.

In the case of the insomniac robins of London and our other big cities, can it be assumed that because the males are singing, the females have had to stay up to listen to them? It's a some-what melancholy thought. In Birmingham, my mother is assured by one of her neighbours that their communal lawn is visited by 'gay robins', and though the suspicion must be that this lady is being deceived by the identical plumage of the sexes, and perhaps also by this same human bias towards the male, her belief raises the valid question of whether the way we interpret bird behav-iour is also blind to the possibility of homosexual partnerships. Whichever way you look at it, there's much that we still don't know, but it seems equally clear that we would do well to begin by becoming more aware of our own unconscious exclusions.

I'm being blown left and right, so I decide to give it up and head for home. As I go I'm watched from the roof of the boarded-up snack kiosk by a crow, hunched morosely, as if it would much rather be squatting on a skull. I commiserate with it aloud, and it eyes me bitterly, then utters a single charred croak before flying off to the wood with weary flaps. A few people are still about, despite the wind, so to avoid them I take the back way. At the top of the alley I stop to admire the blood-red poppies that clot the sides, and my eye follows them all the way down to the sparrow jungle, the town, and beyond that to the ocean itself, where maroon cloud shadows are moored on the mint-and-glaucous waves, marbled white with froth.

After righting the wheelie-bin, I enter the slightly quieter air of the backyard, where the wind is still making local mischief with leaves, stems and a plastic windmill left out by the children. An orange-bottomed bumblebee, either hardy or desperate, loops over the wall and makes a dive for one of our yellow poppies. I feel sorry for the buffeting it's taking, so I bend to hold the stalk for it and am rewarded by the buzzing vibrations travelling down from the happily fumbling creature and up my arm. Just as I open the back door I hear the dunnock start up its thin cascade of notes and I don't need to turn around to know where it's singing from.

Blackcap

5. Songs of Summer

Although it's still only late April, it's a hot, sunny day; the plastic guttering makes little popping noises as it expands in the heat, and the garden drowses under a cloudless sky. The last of the tulips droop and dodder on their long stems, and their rapidly fading glory still attracts an occasional visit from one of the furry-bottomed bees that orbit the terrace gardens on tiny droning motors. In only a month these will run down and return the bees to the earth.

There's still no sign of the virus slowing down. The infection rate is the highest it's ever been, hospitals are warning that their intensive care units are being swamped, and every day the news carries stories about suffering and grieving people, and the miseries of lockdown – both major and minor. Reading these grim reports, we're doubly thankful that we live somewhere so close to nature, and imagine what it would be like to be restricted to a 'recreation area' or a balcony. Not that these places are to be disparaged, of course. Even within this hot backyard, there's plenty to be seen, and there's no doubting the old truth that attendance on microcosms holds interest and brings sensory wealth of a kind usually passed over. Lift a brick to see the slugs, centipedes and gentle armoured woodlice, or, like my children, make earrings of the fuchsia's gory pendants. But when we weary of this?

At this time of year, the immense biochemical detonation is under way that will soon break across our islands in a billowing green wave, foaming in due sequence with the white sprays of first blackthorn, then hawthorn, then elder. The combined energy packed into bulbs, seeds, corms and tubers breaks gently with the force of megatons, and people feel the change and come outside to exchange words with neighbours over the fence or between balconies.

The gradual warming and blossoming of the northern hemisphere brings with it other changes, intimately related and scarcely less profound. Billions of birds, moths and butterflies are on the move, the birds travelling from where they've wintered in southern Europe, the Middle East and Africa to their breeding sites in northern Europe. An advance party of these migrants arrived in late March, or even before, but the great influx really gets going in April and lasts into early May. And each of these birds brings its own songs and calls, that for the next few months will add to the varied chorus of birdsong that had already been rising and expanding in the mild spring air.

Of the approximately 220 species that regularly breed in the British Isles, between a fifth and a quarter migrate here in the spring. Some, such as blackcaps, come from places relatively nearby, such as south-central Europe – still an impressive distance for a bird that could nest in a teacup. Others travel from far away: cuckoos from Central Africa, and swallows from South Africa, more than 6,000 miles away. That swallow darting over the newly cut grass of your local park in mid-April was in mid-March eating flies disturbed by ostriches and antelope.

In earlier times, so it's often maintained, people believed that when swallows disappeared in winter, it was to the bottom of the ponds they skimmed for insects from March to September. There our forebears supposedly pictured the swallows in mud-bound hibernation, after the fashion of frogs. I find the popularity of this rumour suspicious, and more eloquent of our desire to feel modern, to look down on country people and to satisfy our insatiable taste for the 'quaint' than of any plausible belief. What medieval villager, who must surely have lived in much closer relation to the land than we do, and observed its changes much more keenly, could have failed to notice that swallows differ from frogs in possessing wings, and that they gather in late summer before flying away more or less horizontally – certainly not downwards to splash into lakes and pools. And where are the documents of those times that detail the halting emergence in spring, from those same ponds, of the new season's muddy, yawning swallows? No, this is an anecdote that has everything to do with the myth of the credulous yokel, and not much to do with swallows at all.

In fact, migration must have been observed since earliest times. In *De Avibus*, by the sixteenth-century French explorer and polymath Pierre Belon, those with Latin could read about 'passager and migrant Birds . . . who are naturally constituted for distant Habitations, whom no Seas nor Places limit, but in their appointed Seasons will visit us from Greenland and Mount Atlas, and as some think, even from the Antipodes'. Belon travelled widely, and may even have seen birds he would have known from home in their wintering grounds further south. He was one of a leisured elite who had the luxury of literacy, wealth and international travel, but it's not difficult to imagine that the arrival and departure of seasonal visitors was a source of speculation and wonder to enquiring minds among poorer and more land-bound people too – not to mention joy, since their reappearance confirmed that winter's hard recession was over for another year. The anonymous thirteenth-century poet of 'Sumer is icumen in' was not celebrating the arrival of the cuckoo for its own sake, but because it signified that the earth was coming back to life:

Groweþ sed
and bloweþ med
and springþ þe wde nu
Sing cuccu

Seeds are growing
the meadow is blooming
and the spring is made anew
Sing cuckoo!

That birds migrate, then, has long been known, but the question of *how* they do so still occupies scientists today. Studies have shown that many migrants return to the same woodland, park or patch of heath year on year, which is quite a feat of navigation for such diminutive beings. All sorts of explanations have been put forward, and rather than choose from among them it may be wiser to assume that these extraordinary journeys are accomplished by a combination of means. Like mariners of old, birds can locate themselves at times by reckoning their position in relation to the sun and the stars, and it's also thought that they can take soundings from the planet's magnetic field. Terrestrial landmarks help too, and the birds' routes may also be crossed by predictable winds whose dispositions they recognize. The birds communicate mainly by sound, but their sense of smell may also play a part, as some species at least are now believed able to follow odours carried on the wind or rising up from the oceans and landmasses that they fly over.

It's no surprise that migratory birds need to take account both of relatively stable characteristics of climate and also less predictable quirks of local weather. But while nowadays our sophisticated methods of plotting and predicting weather patterns can help us to understand the choices made by migrants, in an age before satellites and computer modelling things were often the other way around, and it was close observation of bird and insect behaviour that enabled early meteorologists to make their tentative predictions. Cuckoos arrive here relatively late, so except in rare years when the spring is retarded by unseasonable snows they're less likely to presage milder days than to set the seal on changes already long noticed. Swallows face the journey a good deal earlier, and are accordingly better harbingers of spring. Those that arrive here before the weather has fully turned prove the truth of the proverb that 'one swallow does not make a summer'.

Successful migrants need to be good weather forecasters for many reasons. Anticyclonic weather often means that skies will be clear, and this is important since thick cloud would obscure the stars they steer by, while the same conditions are also associated with gentler winds. Birdwatchers are delighted whenever an unusual migrant that has been blown off course appears on their patch – but of course it's much less good news for the bird itself. Even if it survives its buffeting, its chances of survival are slight when it arrives exhausted, malnourished and alone. During storms, migrant birds can perish in their thousands, and sailors have often recorded the spectacle of small birds descending en masse to seek shelter on their ships during foul weather. The height at which birds fly is also a factor, with high-flyers being less vulnerable to opportunistic predators but, further removed from terrestrial landmarks, at greater risk of being blown adrift.

Taken together, migrating birds' sensitivity to changes in air pressure, temperature, wind direction and so on constitute an impressive fund of knowledge, without which their journeys would be far more perilous than they already are. Whether this knowledge is adaptable enough for them to survive rapid climate change remains to be seen; one of the few certainties now is that migrants are arriving earlier each year, accentuating the derangement of our seasons just as the medieval poet's cuckoo confirmed their established and more predictable transition.

By whatever means they manage it, our summer visitors come here primarily to find food enough to sustain their breeding, and since breeding is an important motive of song, many species can be heard as soon as they begin to arrive. Some, such as sand and house martins, are not what you'd really call singers; the greater talent belongs to the extended family of birds called warblers. It's just as well that these warblers tend to be good singers, and have such distinctive songs, since many of them are shy of human attention and liable to be seen only as a disturbance of leaves in the heart of a thicket or the greening crown of a tree. Being very small birds, they need to eat a lot to keep going, which means that they also need to keep going in order to eat. They only really keep still when they're singing, so the best way to see them is to patiently retrace the bright path of their song to its source.

The word 'warble' suggests a sound gentle and melodious, but these attributes only apply to the songs of a few of our summer warblers. If you live near fresh water with a stand of reeds, you may be lucky enough to hear sedge and reed warblers. Both have extraordinarily rich and beautiful songs, but they could hardly be described as gentle and melodious – more like sounds you'd expect from a carpentry collective than a choir. Another branch of the warbler family, which includes the whitethroat and the lesser whitethroat, tends to prefer dry and thorny scrub, and these birds' voices are correspondingly dry and somehow tangled, as if the song had become confused in the labyrinth of twigs and freed itself only at the cost of endless backtracking, snags and complications. Meanwhile the grasshopper warbler, as both its English name and its Latin tag *locustella* make clear, doesn't sound like a bird at all, but much more like a cricket, or rather a relay of crickets, since the bird's dry spooling can go on for several minutes at a time.

None of these warblers are common in towns, and for most people in this pandemic spring, when our journeys have been restricted to places we can reach on foot or by bike, our opportunities for hearing warblers have been limited to two or three commoner types, of which the most frequently heard is probably the blackcap. Like the rest of their family, blackcaps are not easy to watch at close quarters, but since a more dapper little bird is hard to imagine, it's well worth the effort to try. Male and female are alike in their grey-brown wings and pale front, but differ in their head colour, only the cock's being black, while that of the hen is a rich liver-brown. As for the song, it's a lucid jangle, aimless but resonant, full and clear, superficially as though a dunnock had been seeing a voice coach, but on further listening almost thrush-like in its fluency and reach. You'll have heard it, I'm sure, and once the songs of the commoner resident species have become familiar, it stands out as the 'something different', the slightly exotic-sounding something that absolutely refuses to extract itself from the middle of the lilac.

Blackcaps are one among many of our summer arrivals that usually make the journey in a single flight. Even willow warblers, coming from Africa, usually fly non-stop, covering a distance of 5,000 miles or more in just a few days. Needless to say, such journeys are immensely draining, so these long-haul specialists eat as much as possible before setting off, even to the extent of doubling their body weight. It's yet to be fully explained why these

warblers don't take breaks; if it seems plausible that early arrivals get to pick the best territories and the fittest mates, then this
should act as a motive for *all* migratory species to get a move on,
yet some species take the journey more easily. Part of the answer
might be that the birds' departure dates are dictated by weather
conditions locally and along the way; there are no plaudits for
even the most extraordinary individual feats of speed and endurance when the object of the exercise is to arrive, not first, but alive
and in a fit condition to sing and breed. Meanwhile, birds that
arrive too early may find that there's insufficient food, and are
then faced with the desperate decision of whether to stay and risk
starvation or backtrack and face likely death from exhaustion.

Unlike warblers, swallows make frequent stops along the
migration route, and may be in transit for several weeks. They
too put on weight before setting off, but in common with other
stopover fliers they also have other strategies to conserve energy
during the journey. Stopping to rest and feed during the day, they
tend to migrate only at night. They probably do this to benefit
from the cooler temperatures and reduced turbulence of nocturnal air, which lessens the risk of dehydration and costs them
less effort in correcting their course when buffeted by winds or
disoriented by sea squalls or Saharan sandstorms.

Given the importance of these birds not squandering energy,
it was long assumed that their migration took place in near
silence; apart from the contact calls necessary to keep abreast
of each other in the darkness, there could surely be no reason
for them to do other than keep their beaks firmly shut, pointed
like compass needles towards the distant north. But while it's
true that the migrants don't sing much, recent observations have
confirmed that more singing takes place than might be thought
expedient for a 30-gram bird travelling thousands of miles
fuelled only by flies. Swallows, warblers, turtle doves and other

summer visitors do give out occasional snatches of song and, intriguingly, they seem to sing more often and for longer the closer they get to their destination. Whether this derives from a hardwired instinct to begin settling matters of territory and matrimony as early as possible, or else from sheer excitement and relief, is another of those questions that everyone is free to answer in their own way.

By the time most of the summer migrants have arrived, the British Isles have been transformed. In the woods by the seashore many tree species are now in early leaf, and ferns are rampant on the rides where cropping ponies hold back the spread of the trees. The blackthorn flowers have fallen or faded, but the hawthorn buds are beginning to unclench in anticipation of their blossoming. Insects are everywhere; skewbald butterflies loop among the path-side flowers, a fly in a ski-mask lands on a flat leaf and rubs its hands together like a cartoon villain, and where a black hollow still holds rain, very new gnats perform very ancient ceilidhs over the stagnant, blossom-strewn water.

Can there be anything more beautiful to northern peoples than woodland in late April? The light seems submarine as it filters down through overlapping leaves of beech, oak and hazel to wag gently on tree trunks or wink from the dark, fleshy surfaces of the holly leaves. Here in the mirror-ball of the wood the light seems almost a material thing; chivvied by mild, ankle-high breezes it scuds along in the leaf-litter and breaks up against twigs into bright little arpeggios of shadow and spangle, runs suddenly up hummocks of newly flowering moss, dithers and then throws itself off in broken beams, beckoning the eye to follow to the open spaces where the fern-banks toss like waves that never come ashore. The maritime effect extends to the canopy, where wind in the beeches makes their leaves hiss like dragged shingle, and more powerful gusts cause branches to plunge momentarily into churning gulfs of fresh green foliage.

Above the dry friction of the leaves and the twittering light, the wood that a month ago rang only with the songs of robin, wren and blackbird is now enriched by new arrivals: garden warbler, willow warbler, blackcap and chiffchaff. Of these species, though, only the first two are certain to have blown in from the south, for over the last few decades more and more blackcaps and chiffchaffs have begun to overwinter here and in other parts of northern Europe. While blackcaps in particular have benefited from people leaving food on bird-tables, this phenomenon seems best explained by global warming. The long, hard winters that were common in the twentieth century are now much rarer, and although the insects favoured by these warblers must still be hard to find during the cold months, there are evidently enough of them – supplemented by seeds and other fare – to shift the balance of risk and convince at least some of the birds that the effort of migration is unnecessary. The urge to migrate may be bred into these birds, but it can seemingly be overridden by changes in practical circumstance.

I could listen to blackcaps all day, but ten minutes with a chiffchaff is already a bit much. The clue to what they sound like is in the name, and they play that two-pitch *chiff-chaff* over and over again. They seem to have drawn the evolutionary short straw when their song is compared with those of their close relatives the wood and willow warblers, both of whom really do warble, and very beautifully too. But it's not just that the chiffchaff has only two notes: the cuckoo, famously, is subject to the same limitation, and the great tit, although capable of a great variety of tunes, seems fond of a bouncing two-tone call usually rendered in bird books as 'tea-cher, tea-cher, tea-cher'. Yet while the cuckoo's call is deep, far-carrying and evocative, and the great tit, pugnacious by nature, sounds off with gusto, the chiffchaff delivers its notes with an earnestness that seems somehow devoid of passion.

Given that its song seems so monotonous and unvarying, it's astonishing how differently the chiffchaff is heard and transcribed in different human languages. To give just a few examples, in Germany they call it a *zilpzalp* and in the Netherlands a *tjiftjaf*, while the Welsh hedge their bets, with *siff saff* and *siff siaff* both in use. These variations may seem inconsequential, but they raise a host of interesting questions. Most obviously, they appear to show that people in different countries hear the same species differently. Cultural relativism can't be entirely assumed, since we know that birds of the same species have different dialects and accents in different places. But if we leave aside the possibility that German chiffchaffs really do sound different to English ones in a way that corresponds exactly to the difference in how they are transcribed in the alphabet, then yes, this does seem to support the notion that how we hear birds is influenced by our own linguistic environment.

As in human politics, controversies about culture tend to colour judgements about moral or aesthetic worth. Who hears and writes the call of the chiffchaff (*zilpzalp*, etc.) more accurately: the English, Germans, Dutch or Welsh? Perhaps it's better to avoid the issue altogether and opt, like the Norwegians, to name the bird after where it likes to sing from, rather than what it (supposedly) says. *Gransanger* (spruce-singer) begs questions of its own, but at least avoids the sort of unresolvable alphabetical dilemmas I mentioned earlier, not to mention the risk of piqued national pride. Those who feared creeping European standardization can continue to revel in the variety of sovereign attempts to hear and write down what *Phylloscopus collybita* really sounds like. Long may we all continue to be inadequate in our own way.

Like many other people, we've chosen to see lockdown as an opportunity to tackle some long-postponed DIY. I'm heading to the superstore in town for a tin of paint, but on the way I can't resist taking a slight diversion to call in on the prison starlings. It's the first time in a while that I've checked on them properly, and the first thing I notice is how smart they're looking; they almost seem different birds from the grubby, smoke-coloured things I last attended to in winter.

One of the by-products of the breeding season is that every adult bird is wearing a bright new suit of feathers. We tend not to notice birds moulting because they mostly do it very discreetly, and with good reason, since the absence of a few flight feathers might make all the difference if a bird is fleeing for its life. It's only to be expected that feathers should become worn and need to be periodically replaced. All birds take care to keep their feathers clean and undamaged, but wear and tear is inevitable, and accidents, narrow escapes from predators and the sheer attrition of weather all reduce their airworthiness and waterproofing.

Birds undergoing the moult can look pretty ragged and half-plucked, so it's understandable that they'd want to avoid doing it in the breeding season, when appearances are most important. Many species, such as blackbirds, moult in the autumn, when territories tend to break down; then they can allow their feathers to fade with the biological imperative to look gorgeous. While resident blackbirds can afford to take their time, though, migratory birds have no such luxury, and this is probably why our small summer vacationers are among the fastest moulters – just a month or so in the case of a willow warbler. That said, there's also variation in how often migrants moult. Chiffchaffs do so just once annually, while summering with us, but the closely related willow warbler changes its feathers twice a year, once after it has finished breeding over here, and again in its African wintering grounds. It's probable that this difference can be explained by the much greater distance flown by the willow warbler, which accordingly experiences greater wear to its feathers and thus needs to replace them more often.

Moulting birds would be unwise to draw attention to the fact by advertising their presence with song, so it isn't only the end of the breeding season that causes many birds to fall silent in late summer, but also their vulnerability while they're replacing their feathers. Some ducks and geese may be so depleted that for a short while they are unable to fly at all, and while this is dangerous enough for them, it would be utterly disastrous for a songbird, which is why their moults are more piecemeal.

It isn't only adult birds that moult. It's during their first winter that most young birds lose their juvenile plumage and grow their first set of mature feathers. I say 'most', because the young of some species, such as robins, stagger their moult so that their body feathers are changed at an earlier stage than their flight feathers. The same is true of starlings such as those I'm watching now; this year's young birds will go into the winter with mostly adult feathers except for those on their heads, which will retain their ash-brown juvenile colour until next spring.

But all this is still months away. At this time of the year no bird has a thought for the moult, being too busy showing off. On my way down here, passing through the park, I saw that each cock sparrow wears a spotless black cravat, the moorhen has thrown on a surprising cape of magnificent liquid bronze – like the slick on a fresh cow-pat – and even the feral pigeons look dressed for success, their eyes polished to amber, a flash of iridescent green and violet at the neck.

But amid all this splendour of colours and hormonal enrichment of tones it's the starlings that stand out most of all, and I'm glad that I've stopped by to watch them here in this odd zone between the prison, a barracks and an allotment, which is where this particular extended family chooses to base itself in spring and summer. During the winter, they roost in some numbers on the superstore roof, taking advantage both of the extra degree or two of warmth provided by the building's heating systems and the odd windfall of crumbs from an impatiently severed baguette that won't fit whole into a shopping bag. By March they extend their forays into the miniature forests of daisy, dandelion and clover on the car-park verge and the waste ground of the building site, walking fast but stopping often to peck among the leaves for the newly abundant insects. Although good for feeding, though, there are too few nesting sites there, which is why they are reliably found here by the prison a few weeks later. For humans this part of town may be slightly forbidding, but for the starlings it has it all.

Between the ages of eight and ten, I spent a lot of time as a bird of prey. I'd glide around the house with stiffly outstretched arms, and when I spotted a potential meal – almost always my sister – I'd tuck my arms behind my back to imitate the folded wings of a stooping falcon and speed towards my target, who would usually respond by kicking me in the guts. As my knowledge of predatory birds grew, I incorporated it into my role-play so that, for example, when riding my bike as a sparrowhawk I mimicked with my pedalling action the characteristic 'flap, flap, glide' flight of that bird, and on hot summer afternoons I circled the top end of our cul-de-sac, trying, like a buzzard riding a thermal, not to 'flap' at all.

Although by this age I was interested in birds of all kinds, it was birds of prey that dominated my imagination, and I've since witnessed the same preference in other children, including my own. I doubt that this is coincidence, and I ascribe it to a developing desire for the self-sufficiency and wild freedom that raptors seem to represent for people. Children go through alternating phases of seeking comfort and then independence from their parents, and birds of prey seem to be an ideal embodiment of the latter – a thrilling yet safe form for shape-shifting children to inhabit while they explore the limits of their own boldness.

If I'm right about this, it might also help explain why, as I age, I'm increasingly aware of being drawn to more sociable birds. Not that the excitement of watching raptors has diminished for me – not at all! If I'm lucky enough to see a sparrowhawk or peregrine then I still experience the old childhood thrill in their imagined qualities of autonomy and confidence. But although my reaction to the birds is the same, my desire for autonomy has probably lessened, or at least been balanced by a mature appreciation for the society of like-minded people. The birds I most like to watch nowadays are those gregarious species that also seem tolerant, or even fond, of human company. Among urban birds this tends to mean jackdaws, sparrows and especially starlings.

Starlings are forever busy, and as I look at them now, it's hard to know where to turn first. For a start, there's a constant traffic of birds over the high walls of both prison and barracks into the usually quiet yards for exercise and drill where their nests will be safe from predators and frequent human disturbance. The quiffs of razor wire that top the barracks wall are an appropriate vantage for the starling on sentry duty, and he or she oversees with a vigilant eye the business of the other birds, which at the moment is concentrated in the playground that adjoins the allotment. A month into the pandemic, the playground gates are still locked shut behind ugly plastic tape, and with no children to pound it down the grass has put up hundreds of dandelions, both the shaggy saffron flowers and the fluffy grey 'clocks'.

I grew up with starlings. In the 1980s, Birmingham city centre was a winter roost for tens of thousands of them, and the birds coated the narrow ledges and sills of the buildings an inch thick with their white droppings. On evening car rides into town I watched from the back seat in utter awe as their pulsing flocks whirled around the sky, causing the city lights to flicker as they passed in front of them. Though it seems almost impossible to me now, there were without exaggeration so many starlings that the sky was grainy-black with them, and they swirled and pulsed in the sky-space between the concrete towers as though a dark dough of poppyseeds was being stretched and kneaded by invisible hands. But those birds – perhaps half a million of them in hard winters such as we sometimes had back then – were much more than a visual spectacle, because their noise was just incredible. Starlings have arguably the greatest repertoire of any British bird, and even an individual can fool you, by its chatters, pop-gun detonations and saucy whistles, into thinking you're in the presence of half a dozen. Imagine, then, the racket of half a million starlings, completely eclipsing even the bellowing of the Bull Street drunks, streaming into the night air above the city like a swelling tide of whirring and clicking that could be heard a mile away.

I get a faint echo of all this as, leaning against the allotment fence, I take in the action in the playground, temporarily converted to a starling canteen. The birds sortie in pairs and small parties, and as long as I keep still they're confident enough to let me see the constantly shifting richness of their plumage. Glossy black from a distance, when seen in close-up each bird is its own galaxy of spangles on an iridescent ground of indigo, bottle-green and metallic blue – every colour of the petroleum rainbow. Add to this a pair of jaunty pink legs and a multi-purpose omnivore's bill of palest ivory yellow, and you have a gorgeous bird indeed. The four starlings nearest me have their heads down nearly all the time, nodding left and right in time with their somewhat clockwork carriage, glancing and pecking all the while. Gregarious birds can afford to spend more time looking down at the ground, knowing that their associates are watching their backs. In just a few weeks the playground weeds have grown extraordinarily high, and sometimes the birds disappear altogether, only for a dark head to bob up a few seconds later above the vegetation.

The starlings aren't the only birds drawn to this place. Sparrows are fussing in the thorn hedge, and they constantly run small missions to a weed patch by the barracks wall, standing erect and stretching their legs and necks to peck off dandelion and groundsel seeds for nest material, before sinking back into their usual dumpy posture and then departing with a fluff-muffled chirp. A collared dove is also busy on the ground, but it flies off with a bleating cry as a woman goes by with a pushchair, its splayed tail feathers translucent for a moment as it ascends into a sunlit cloud, looking for all the world like a dove on a piece of devotional junk mail. A flock of feral pigeons fly laps of the square, sometimes passing close enough for me to hear the rush of air from their wings, while further away a wood pigeon in full-throated *basso profundo* expresses its passion or ire – I can't tell which.

But for me, it's the starlings who steal the show. Though it's still diligently swivelling its pointed head to scan in turn the sky and ground, the sentry on the barracks wire seems to be enjoying its work more than is appropriate, or at least this is the impression created by the salvos of chuckling and naughty whistles that it's giving off. Suddenly it makes two short, sharp warning notes, and the birds on the ground instantly scatter – but it must have been a false alarm, as within seconds they're prodding among the grass-stems again as though nothing had happened, swapping small talk with each other as they go over the ground.

On top of the allotment fence, one adult starling of uncertain occupation – but probably a poet – is in full extraordinary flow. Extending his head to the sky, he comes out with a bizarre sequence of near-simultaneous pops, gobbles and wicked cackles. I can see his throat working away under its gorget of glossy feathers, ten to the dozen, as if he were manipulating at least half a dozen syrinxes to produce this effect, which sounds like nothing so much as a bag of half-wound music boxes rolling down a hill. None of the other birds seem to regard this as anything out of the ordinary, but when he finally ends the performance on an impish descant whistle it's only the sudden appearance of a skateboarder that prevents me from erupting into spontaneous applause. Such virtuoso displays are commonplace for starlings, and the only question left unanswered is how they can produce all of these ridiculous noises yet still keep such a very, *very* straight face.

These sparrows and starlings are prospecting for nest sites, if they're not building already, and it might be thought that the arrival of the summer migrants would be an unwelcome source of competition for the choicest places, yet this seems not to be the case: these migrants are no more likely to come into conflict with robins and wrens than the native birds are with each other. There isn't always an overlap between the preferences of the visitors and our resident birds. Some spring migrants make nests quite unlike those of the natives, such as the holes in the soft soil of riverbanks beloved of sand martins, or the precarious muddy hemispheres that their cousins the house martins somehow gum to the underside of eaves. Arriving a little later, perhaps not until early May, the nightjar makes do with a flattened patch of bracken; meanwhile the cuckoo famously declines to build at all and instead lays its eggs in the nests of other birds. The summer visitors more likely to be seen in towns – chiffchaffs and blackcaps – make nests in more conventional places, but there's still plenty of grass and twigs to go round.

In terms of song, too, by late April there can be heard an agreeable blend of familiar voices and those recalled from the previous summer, and the dawn chorus – earlier every day now – may consist even in the most unpromising inner-city areas of the intermingled voices of ten or more species, all welcoming the sunrise in their own way, and each intent on pairing and finding suitable places to raise their young. But we can't take these summer visitors for granted, for they're in serious trouble. Almost all the British nesting species in most rapid decline are those that visit from southern Europe and Africa. Many of the species that would have defined the start of summer even a generation ago are either absent altogether or so depleted for it to be a case of 'out of sound, out of mind'. My grandparents still heard corncrakes and quail rasping in the fields behind their house in the English Midlands, and my parents when they were young could have heard nightingales and turtle doves over much of the country. I've never heard any of these species in Britain, though they're still fairly plentiful on the continent – incongruously, I've never heard so many nightingales as in Moscow, in the grounds of the Kremlin.

That the decline of these species is an ecological problem is hardly worth saying, but in ways that are just as real, yet more difficult to articulate and measure, it's also a great loss for our sense of who we are as human beings. It's beginning to be understood that the disruption to the seasonal cycle caused by climate change is a cause of both conscious and subliminal disorientation and distress. In the British Isles this is most obvious in the shrinking of our winters – but the loss of other seasonal markers also reduces the variety and enchantment of the world that is such a source of intrinsic delight but also helps us to locate ourselves in time. We might not yet be desperate to the same degree as the Icelanders, whose mental health is being affected by the loss of the glaciers, but our basic predicament is the same.

How do you know it's winter if it's 15 degrees and drizzling on Christmas Day, just as it was on the first day of the autumn term? That's easy: the shops are full of cards showing robins perched on snowy branches. Not only this: if you go for a walk in the park you may hear the wet chuckle of fieldfares in the trees, and if you put the bins out at dusk then there's a good chance that you'll hear the whistle of redwing on the move, keeping in touch with each other in the darkness between stars. But when it gets to 15 degrees in Finland too, and the redwing and fieldfare don't need to come here anymore, then we'll only have the fiction of the Christmas cards.

So it is also with our summer birds. The cuckoo no longer comes to St Brynach's cross, so how would we know that it's April? And even beyond this, the songs of summer don't just mark the seasons: in a subtle way they *are* the seasons. I spent part of last summer in central Europe, and on my walks I often heard turtle doves purling from the wooded hills. The song of a turtle dove is like the aural equivalent of a heat-haze, the gentlest corrugation of air, always just on the edge of your hearing. As I listened, it struck me that for generations in Britain before me, this song, along with those of the corncrake, cuckoo and quail, probably distilled the very essence of what summer was; summer was experienced as summer, not only because it was warmer and the football stopped, but also because of this sound, now little more than a rapidly fading memory.

The reasons behind the decline in our summer birds are multiple and overlapping, and the loss of the migrant species needs to be seen in the wider context that there are millions fewer birds in Britain now than there were when I was a child – but there's no doubt that habitat depletion at every stage of their journeys plays a part, as do trapping and hunting and the catastrophic decline in insects. Most of our summer birds are insectivorous, and none more so than my favourite of all: the swift. I've loved these birds since I was a child, and though my wonder only grew when I learned that they feed, mate and even sleep in flight, the immediacy of my awe was in the mere sight of their dusky crescent bodies ripping through the air above the house, as if shards from some long-exploded moon had just shrieked into our atmosphere. On muggy summer evenings, either side of thunderstorms, they would descend from the higher air to hawk insects and chase each other in small posses between the chimney stacks. They were seemingly so exhilarated by their own pace and dexterity that they couldn't help but emit those thin screeches, hardly melodious in themselves, which nevertheless became as joyful a confirmation of summer as the pleasanter calls of the warblers. But for years now I've watched the swifts grow fewer, and in a grim inversion of the happy expectancy of summer visitors that ought still to be normal, I'm aware that on some level I'm already steeling myself for a time when I may no longer hear them at all.

One morning in mid-May I stop to watch a single swallow skim the park lake. Almost too quickly for my mind to trace, it loops the island trees and returns three or four times to take a mouthful of water, pocking the dark water white with ripples and spilt droplets before rising again sharply to show its anchor-shaped silhouette against the sunlit clouds. The anchor image makes me think again of the swallow's hard passage over the undrinkable water of the ocean, and of the long habit of sailors who to this day sometimes tattoo a swallow into their skins, an echo of the former belief that swallows are the heaven-bound souls of the drowned. A different tradition symbolizes the respect that mariners have for the swallows' own feats of endurance and navigation: for each 5,000 nautical miles that a sailor travels, he earns the right to a new swallow tattoo.

Throughout history we have superimposed our own meanings on to birds, often at the cost of seeing them as fully independent forms of life. It may even be that the biological and symbolic viewpoints are incompatible, so that we're forever condemned to value birds either as essentially meaningless organisms such as we too would be without culture, or else as embodiments of human fears and aspirations, but never both at once. If I'm right about this, then there are also two different but related forms of comfort to be derived from the sight of a swallow during this period of lockdown. As winged embodiments of the freed soul they offer consolation to those immured physically by quarantine or emotionally by depression or grief; but if we peel off all symbolism and make the effort to see them only as living organisms, then the birds' very autonomy, the fact that they don't need us, can also be a source of deep comfort. This goes not only for the *sight* of birds but also or perhaps especially for their songs. As Edward Thomas expressed it, 'Beautiful as the notes are for their quality and order, it is their inhumanity that gives them their utmost fascination, the mysterious sense which they bear to us that the earth is something more than a human estate.'

Jackdaw

6. Jackdaws in the Chimney

As though under enchantment, the walls and trees of the town have begun to speak. From brick crevices, cracked masonry, shadow-hatched thorn thickets, breezy green mezzanines and fissures in tree trunks comes a seeping and a wheedling of blind, hungry birds. The movements of the adults, never as random as we usually think them, are now entirely governed by the quest for food, and anchored to the nest site by invisible threads. The days are long, and insects and seeds are plentiful. Wake at 5 am, stay out until dusk – you will never find the nests unattended, nor see the little creatures rest.

Birds' nests are not difficult to find, though it seemed so before I began to pay attention to their songs, alarm calls and routines. Now each morning I do a round of the nests that I've discovered, and thus I have a routine too, my own orbit altered by the great gravity of their purpose in preparing for the world the next generation of birds. My first stop is at the nest of a pair of collared doves, who have built alarmingly low down in a pollarded linden outside a drug rehabilitation centre on a quiet street. The nest is fairly well screened by leaves, but there's a passage for the eye into the cool shade near the trunk, at the other end of which I can see the mild black eye of the dove, set dead centre in the putty-coloured feathers of her head. I don't want to draw attention to the nest by lingering, but for the five seconds I dare to spend standing still, she regards me in silence and without movement. Are we recognizable enough objects to doves for them to feel the pressure of our gaze? Somewhere further down the road her mate gives the peevish, trembling cry that they often make in flight, and I move away quickly.

From the doves' nest to the park is a short walk through empty streets, under a tangle of song. At the park's centre is a strange extrusion of jumbled rocks, partly concealed by rhododendrons and a great willow, and it's in a crack at the very bottom that a great tit pair has chosen to nest. When I found it, my first thought was that I hoped the birds knew what they were doing. The nest is at ankle height, or more pertinently, about level with the head of a questing weasel. Surely they must live in the park, and if they do, this nest is defenceless, for a weasel is just a lean, toothed muscle, and there's no crack wide enough for a great tit to pass through that won't also admit this insatiable little hunter. For now, though, importunate squeaks are still emerging from the wall, so there's nothing to do but trust to the parent birds' judgement.

Nearby is a coal tit's nest, which, although also alarmingly low down in an alder trunk, is at least screened by nettles; there's a nuthatch nest, too – in an oak tree hole clearly once excavated by woodpeckers. Nuthatches are unduly noisy little birds, and this pair make no effort to approach the nest site with discretion, both parents uttering their strident *weet* call before disappearing into the cavity with bills full of insect larvae and spiders. The wood being so quiet, I can even hear the scritting of their long toes on the bark as they re-emerge, sometimes descending the trunk head-first as though this were the most straightforward and sensible thing in the world.

These days, we're discouraged from taking any more than a distanced interest in birds' nests, on the good grounds that any disturbance, however well motivated, might cause the parent birds to abandon them. Against this, it's possible to argue that the end of 'bird nesting' removed a form of intimacy with the natural world that in many cases fostered the knowledge and enthusiasm of future naturalists. Many reminiscences of early twentieth-century childhoods include episodes of nests being sought and the eggs examined or taken, but the practice is of much greater antiquity than that. Tacked to the wall at home I have a postcard of an artwork by Pieter Brueghel the Elder, painted in 1568, that shows a man climbing a tree to rob a nest of its eggs, and though this particular painting is usually inter- preted allegorically, the moral point would not have been taken had the real habit of nest-robbing not been very common. On the other hand, there have always been those who pleaded for nests to be left alone, such as the speaker in a Scottish song called 'Thou Bonnie Wood o Craigielea', written by Robert Tannahill some time around the turn of the nineteenth century, who begs would-be destroyers to think of the beautiful birdsong that they are depriving themselves of:

Awa, ye thochtless, murd'rin gang,
Wha tear the nestlin's ere they flee!
They'll sing you yet a cantie sang,
Then, oh! in pity let them be!

Happily, not all those who seek out birds' nests are so destruc- tive, and many explorers of hedgerows and fields would have been attracted simply by the inherent interest of the birds and eggs themselves. The doyen of inquisitive nest-seekers is the poet John Clare, who, a generation after Tannahill, wrote many precise and affectionate verses about his encounters with nests in the Northamptonshire countryside. In 'The Wood Pigeon's Nest', Clare confesses that 'in a curious mood I've oft been led/ To climb the twig-surrounded trunk and there/On some few bits of sticks two white eggs lie'. Some nests, like that of the raven, are built so high as to be safe from the attentions of the village boys who challenge one another to reach it. Not so fortunate is the owls' tree-hole nest described in 'The Martin', which tempts 'gipseys often and birdnesting boys' to climb the trunk to reach inside it; but though the climbers get close enough to 'Look in the hole and hear a hissing noise', the parent owl is able to defend the nest and ends up driving the intruders back down the trunk.

It's characteristic of Clare's exactness to include the 'hissing' of the owlets, and in another poem he uses this same word to describe the sound made by the wryneck in defence of its nest. The repetition of 'hissing' shows no want of attention or vocabulary on Clare's part, but quite the opposite: the young of many bird species do indeed give a dry hiss if they're alarmed, something I used to notice when I went too near the nest box of the blue tits in my grandparents' garden. It's been proposed that this sound repels invaders because of its similarity to the papyrus rasp of an adder, and in the case of the wryneck this might be true, since Clare's account of the disturbed adult waving 'her head in terror to and fro' also suggests the movements of a snake. Whether or not any imitation is at work, a hissing sound seems universally threatening, or else it wouldn't be employed by birds as dissimilar as tits and swans.

We rise early with the children in this sunniest of springs, and from the moment we push wide the windows we can hear the muffled morning whickering of next door's jackdaws, as the adults and young birds converse within the chimney. Eight chimney pots along the road contain perfunctory nests of twigs, the birds favouring chimneys capped with rain cowls low enough to admit themselves but exclude the larger crows that would take their eggs and nestlings. From within these chimneys, and from

the parent birds standing guard on the ridge tiles or flying in
with takeaway grubs, comes a near-constant chatter of the most
delightful sounds. I suspect that not even to other jackdaws does
this resemble what we'd call singing, but if song is a form of
emotional communication then we need to widen our definitions
and put jackdaws on a par with nightingales for the richness of
their vocabulary.

Jackdaws are often seen in the company of other crows, and
they even co-nest and co-roost with rooks, but there's little risk
of confusing them by their calls. True enough, they all sound
distinctively crowy; they share a family resemblance in the same
way that pencils of different hardness are kindred graphite. But
while the call of the common or carrion crow is never gradable
as less than a severe 8B, and rook-calls are a sleepy and smudgy
HB, jackdaws are somewhere in between, being capable of
both a high, hard clatter and a range of softer tones according
to their mood and situation. Even the characteristic *jack!* that
might confer their English name can vary greatly: while usually,
and especially in flight, it sounds challenging and affronted, the
same sound uttered more quietly at the nest site, or conversation-
ally while feeding, has an affable mood unique among British
birds.

Listening to them now, as a warm wind from the sea rises up
our narrow gorge of concrete, I feel a great affection for these
birds that have chosen to share our houses. One parent bird is
on sentry-duty on the chimney pot next door, while below its
mate walks down the road alongside the kerb with an odd, wide-
legged and high-stepping gait, as though its shoelaces had come
undone. This jackdaw stops frequently to peck at ants or crumbs
or goodness knows what other meagre food can be found on the
roadside. Although they'll always keep an eye on you, jackdaws
are not very shy of people, and especially in the confidence of

numbers they may approach near enough to be fully admired for their black and silver plumage and especially for their eyes, which in shade seem pale smoky grey but are revealed in sunlight to be the most delicate imaginable shade of blue.

Something disturbs the bird on the road, or perhaps it has a morsel for its chick. At any rate, as it flaps up quickly from the kerb it emits a sharp, clear yap, which is immediately answered from inside the chimney by a call which is recognizably the same but higher-pitched yet also more opaque and gauzy – although this could well be just the muffling effect of the brickwork and cowl.

I imagine the chicks crouching within the warmth, stink and semi-darkness of the flue. Jackdaws hatch from pewter-blue and chocolate-mottled eggs, emerging blind and lizard-like on to a rough assemblage of sticks softened somewhat by long grasses and horse hairs if it's a countryside residence, or more likely weed-stalks, seed-fluff and even wisps of pet hair if the birds are urbanites. Their first feathers begin to grow almost at once, and within three or four weeks the chicks already resemble miniature adults, albeit with the oversized heads and wobbly carriage of infants everywhere. Clutches of four to six are usual, and may even rise to seven or eight. Since the eggs don't all hatch at the same time, the chicks' chances of survival diminish the more older siblings they have to compete with them for food.

These chimney-dwellers aren't the only nesting jackdaws I've found this season. Half a mile away, in the park, there's a jackdaws' nest in a sycamore trunk where disease or a gale has broken off a branch, leaving a dark, rough-edged hole to mark the amputation. Here a jackdaw pair have reared two young ones that, to judge by their protruding heads and importunate clamour, aren't far from fledging. This nest being much less accessible than that of the doves, I don't scruple to stand on the pavement beside the railings and crane up to where the young ones peer back at me with inquisitive twisted heads. Jackdaws, and crows generally, are highly intelligent birds, one of a select group that can recognize individual human beings. My grandfather Williams befriended a semi-tame jackdaw that would come to him for food and even follow him to his shifts as a bus driver, yet withheld the same favours from anybody else.

Later, towards dusk, the terrace birds and those from the park will join forces and descend to the newly mown playing fields with a great noise – not cackling but *crackling*, a bit like car tyres rolling over beech mast. While they're there, plundering for grubs and seeds, the young will fall silent within their chimney nests at about the same time that my brood also go to their beds.

When bird books mention the 'nesting season', the implicit reference is often to the relatively brief period when birds are actually sitting on their eggs or feeding their nestlings, but for the birds themselves this is just the middle phase of a much longer process. Blackbirds are among several species whose nesting season begins as early as February with the quest for suitable sites; both male and female blackbirds take part, but the hen invariably has the final say. The pretence of equal labour is abandoned altogether for the actual building of the nest; the hen works on this alone, and if the cock does occasionally peck at a bit of nesting material, this is so half-hearted as to suggest an involuntary tic more than any real intention to help. In fairness to the male birds, they are usually diligent both in keeping a close eye on the nest and also, when the time comes, in helping to feed the nestlings.

But what an absurd thing to begin life encased in an egg! So it seems to us, anyway, and not only because almost all mammals do things otherwise but surely also because the only time most modern people see eggs is on supermarket shelves or in frying pans. So not absurd then, just a question of seeing matters differently, our own habits included. But nevertheless, how unlikely it seems that a being which in April was folded up inside the almost-darkness of a three-inch egg is by May flapping around in the bright sunshine.

Not that the egg-bound chick is entirely passive; though its range of vision and movement are obviously very limited, it still has its voice. While not even the most precocious bird has ever *sung* from the egg, unhatched chicks can communicate with their parents by cheeps and squeaks, and in this way establish a bond that will last until they achieve independence many weeks later. This is vitally important in the case of colony-nesting seabirds such as puffins, terns and guillemots, since amongst all that caterwauling a chick needs to be able to recognize its parent, and vice-versa, as soon as it has hatched.

But parents and egg-bound chicks don't only call in order to recognize each other. Astonishing though it might seem, adults 'teach' their eggs the meaning of various different songs and calls, so that when the chick has hatched it already knows how to respond. This must play a role in developing the young bird's sense of what it is – that it is an individual of this species rather than some other. (If the ugly duckling's parents had been more attentive, it would have known from the off that it was a swan, thus forestalling its own teenage angst and also sparing generations of human offspring a dubious morality tale.)

Alarm calls, too, are transmitted from the adult to the immured chick, along with the stress hormones that accompany them. Eggs exposed to their parents' calls of fear and distress discharge more nervous chicks than those of less stressed parents; the kinds of interaction that take place between birds and their unhatched offspring are in most respects functionally identical to those arranged for in mammals by the chemical exchanges mediated by the placenta. The one big difference as far as I can see is that, direct placental care being impossible for birds, sounds must take on a greater burden of communication, which may be another explanation for why birds, of all vertebrates, are the vocalists *extraordinaires*.

Since birds come in such a great variety of sizes, it makes sense that their nests do too, from the espresso-cup of the kinglet, carefully woven of moss and spider webs, to the huge penthouses of herons, consisting of metre-long branches which look as if they've been thrown together anyhow. You'd expect herons to nest in the reeds, like swans, and it comes as a surprise to see these tall, stiff and severe-looking birds building their nests communally near the tops of tall trees. These heronries, accommodating anything from a score to a hundred birds, are often ancestral; left alone, the birds tend to use them year on year. But a combination of iffy workmanship and autumn gales takes its toll; each spring the birds are obliged to make renovations, and it's at this time that things tend to get noisy. I've been lucky enough to live near two heronries, and watching this spring-cleaning might be my favourite birdwatching experience of all.

The refurbishment of a heronry is conducted with great excitement and ceremony. Every time one of a pair flies in with a stick, it celebrates with a *kraak* or *gronk*, seemingly produced in the throat, so not seriously muted by the stick. As soon as the incomer is heard or spotted, its mate lets out an answering croak; by the time all the neighbours have joined in, the whole riverbank is loud with sounds properly indescribable but vaguely reminiscent of something amiss in the plumbing. Along with everything else, beasts and birds of prey for a mile around must be aware of the noise, but that's not a concern for a bird 70 centimetres tall and with a bill like a broadsword. This might explain why heronries are sometimes surrounded by satellite towns consisting of the nests of other, more vulnerable birds, such as jackdaws. A sparrowhawk or peregrine falcon in pursuit of a jackdaw would sooner pull up short than head into a heronry.

As well as making their guttural croaks, excited herons sometimes clap the upper and lower halves of their beaks together in a hollow clattering of cartilage. This usually happens when the stick-bringer has made it to the nest, and the courtly ritual of bill-clapping is accompanied by many anguine movements of the neck, and much waving of crested heads. All this happens around April, and it's the only time of the year, except perhaps when they're being mobbed by crows, that you'll witness herons being truly animated. Later in the year, after the eggs have been laid, but also when their spirits are dampened by summer showers, the herons take up their familiar hunched demeanours, as if a convention of dour deputy headmasters had by some oversight been booked halfway up a Scots Pine.

After their eggs have hatched, the birds continue to fly in with whatever fish, frogs, mice and smaller birds have found themselves on the wrong end of those indiscriminate beaks. Arguably it's too stiff to be truly graceful, but the flight of a heron impresses by the sheer breadth of its span and the dark, doomy portico in which its wings are held. Only once in my life did I see a heronry spooked, and it was as though a viaduct had been detonated in slow motion, each double arch of the disassembled structure flying off on its own, either side of a trailing pair of legs and a long, folded neck.

Two months into the pandemic, the mandated hour's exercise feels more and more limiting and claustrophobic. Most days I walk the same paths and exchange distant pleasantries with the same people. It's starting to feel as though our very movements are prescribed, and in an effort to shake off this feeling I decide to take a different way, down towards the university campus, where perhaps I'll come across other nests. It's the first time I've been down here since lockdown began, and it's very odd to see the place deserted. I assume that most of the international students have somehow made it home; only the small student shop remains open, and on the concrete parade, which would usually be so lively, the sole human noise is the echo of a distant mowing machine from the park. But while things are strangely quiet at ground level, up on the roofs of the buildings it's clamorous business as usual.

In a town like this, which has few tall buildings, those that do exist are greatly in demand among the large population of seagulls. The big department store in the centre is much favoured, but most desirable of all are these university buildings, which to the gulls are irrelevant as seats of learning, but highly prized simply as seats. From the top of these towers the gulls – mostly lesser black-backed, but some herring gulls too – can survey both the ocean and also the park and student dorms that in normal summers provide most of their food. Have you ever seen a herring gull eat a herring? I doubt it. But as with the ancient human belief that everything on land has its equivalent in the sea – aquatic cities populated by mermen and mermaids, and so on – herring gulls live in a world of analogy in which the science block is a cliff, a bin lorry is a trawler and a kebab is a herring. How badly the birds are faring in the absence of students may be judged by the sight of scores of them morosely trampling the grass of the playing field in an attempt to conjure worms to within range of their beaks. Taking the longer view, though, this is likely to be the healthiest generation of gull chicks in quite some time.

For now, in late spring, the tower blocks aren't just look-out posts and places to rest, but also hatcheries, and each individual nest is defended with all the honking and chuntering that these birds are capable of. The nests themselves aren't much to look at, being nothing more than artless heaps of twigs and straw. But in the parallel universe of the gulls each nest must be a thing of rare beauty; not only is it defended with the utmost vigour but it also causes paroxysms of excitement between the proud builders, manifest to the rest of the world in a captivating sequence of amorous gestures, from neck stretching and bill kissing to maniacal gobbling. This pair-bonding is noisy enough, but it's nothing to what happens when a strange gull overflies their nest or lands too close. And when a blocked gutter obliges some unlucky member of the university maintenance staff to ascend to the roof, the sound is like nothing on earth. Pushing up the skylight must be like opening the gates of hell.

Yet this only goes to show that these much-maligned birds are commendably vigilant parents, and though anyone who's had a gulls' nest on their chimney will confess to the sleepless nights spent on firearms websites, the flip side is that audio compilations against insomnia often feature the gentle crying of distant gulls. Though we might not always admit it, we islanders find something irresistibly evocative in the sound of these birds.

Under the hostile stare of the gulls I leave the precinct and head back to the park, following a seldom-trodden track that skirts the trees. On this margin between grassland and wood are many saplings of sycamore and rowan, and the latter already show green berries clustered beneath their long, slim leaves. Soon the berries will warm to olive, although it will be late autumn before they take on the scarlet colour that will attract both native and wintering thrushes to gorge themselves. Here too are many bramble bushes, and as I walk I take note of where to bring the children blackberrying in August. The grass is growing long now, and vortices of flattened stalks show where mammals from shrews to foxes have their runs to and from the wood. Somewhere off the path I hear the single anvil strike of a woodpecker, so I turn aside, bending slightly to enter a sort of natural chamber made by the overhanging branches of some mature oaks.

I wasn't to know that this was where the wrens would have their nest. One second there's just the barking of a distant dog and the soughing of wind in the leaves, the next instant it's as if I've sprung a boobytrap of alarm clocks. It seems that hatching has only just finished, as the four chicks, gathered around an old split holly stump, flat-headed and streaky, seem not to know how to use their wings. They look too bewildered to do much but squat down like feathered frogs and hope I go away – but the parents aren't leaving this to chance. No less protective of their young than the gulls were, they come very close to berate me, straining their three inches upwards for maximum intimidation, bobbing in fury and churring wildly all the while. And it works. Despite the great size difference between us, something about the birds' vehemence really is off-putting, and I back off, feeling chastened.

Some birds' nests are impossible to miss, while others are rarely seen. Generally speaking, the larger the bird, the more conspicuous its nest – and because they crowd together, colonial nesters tend to be the most noticeable of all. Right on the edge of town there's a straggle of rooks' nests spread between five or six beech trees on a windy ridge. Although they do sometimes come on foraging raids into town, it's still generally true to describe rooks as the country relatives of our more familiar crows, from which they can be distinguished at most times of the year by their cinder-grey beaks, sloping foreheads and rather bandy-legged gait. It's as if all that time spent swaying in the canopies of trees leads them to over-compensate when on solid ground, giving them the 'sea-legs' of sailors. The best way to tell apart these species, though, isn't by their looks but by how they act – for while crows are usually seen singly or in pairs, rooks are highly sociable, feeding, roosting and nesting in rookeries that may house hundreds of birds.

If pressed for my favourite British birdsong, I might opt for that of the rook. True, individual rooks can hardly be said to be beautiful singers, but the corporate hubbub of a summer rookery is the natural sound I'd carry with me into exile. Rooks tend to return to the same rookeries year after year, and some of their ancestral pads are old indeed. Charlotte Brontë refers to a rookery several times in *Jane Eyre*, and it's surely possible that the same colony of nests that still exists within earshot of Haworth Rectory, where Brontë was raised, inspired those passages. Though rook nests are hard-wearing, they're still vulnerable to winter storms, and what serves as a roost may be too precarious to entrust with eggs. As with the herons, early spring is a time for renovation, and though rooks are not migratory, this renewed activity at the colony is in some countries, such as Russia and Ukraine, welcomed along with the arrival of swallows and cuckoos as a sign of spring.

It's during these bustling weeks of spring cleaning that rookeries are richest in sounds, from the harsh warning alarums of the sentries, and the amiable bickering of birds barging past each other with twigs, to what seem like exclamations of either approval or admonishment when the cock bird presents his offering of nest material for the inspection of the hen. The gossiping quality of rook language makes it sound very human, and despite the occasional squabble or more serious flare-up of violence, the rooks seem to take an active pleasure in community that extends far beyond its evolutionary practicality. If it's possible to reduce the babel of a rookery into one emotional alloy, then for me it would be contentment. The cheerful language of rooks never fails to rub off on me, and listening to them brings me the same easy pleasure to be found in being a stranger among speakers of another language: the effort to understand can be rewarding, but at the same time it's also enjoyable to let go of meaning, and simply let the sounds and rhythms wash past you without interrupting the passage of your own thought.

As spring turns into summer and the fine weather persists, the parks begin to fill with people once more, although most are still at pains to avoid each other. When I'm out with the children I'm always having to steer them to one side of the path if there's someone coming the other way. But when I go out alone, I can pay more attention to the changes that are taking place in the landscape. The trees are almost at the height of their powers. It's hot, but coolness rises from the dark water of the narrow stream that winds in small meanders where it's flowed aside to avoid a swollen root system or to eat into the softer gravels of the bank. As usual, I stop for a while to watch the water drag its clear skin over the stones. How beautiful they look when wet, and how dismally ordinary when they've dried out in your pocket! On either side of the stream, flat leaves are helipads for resting flies. The nettles are now growing fast; a few have overbalanced and fallen from one shore to the other, forming hairy gangplanks for caterpillars and ants. I can't actually see any caterpillars on these nettles, but the yellow ragwort behind me is festooned with them – the black and yellow larvae of the cinnabar moth. If I come too close they writhe like possessed pipe-cleaners, and some even drop into the long grass in their haste to flee.

I follow the stream to the park lake, where a warm breeze pushes little rafts of alder seed and feathers over the shallow water. There's a bench here, under the grizzled alders and adjacent to the weed-ridden bowling green with its boarded-up pavilion, which even in its dilapidation still transmits a sepia aura of gentility. Watching water is good for weary eyes; I'd like to unfocus completely, but always life compels the attention. The reflections of gulls plane the dark surface, but break up in the wake of a moorhen that scoots across the water in coordinated jerks of its head and legs. The effect is briefly comical, as though the body were condemned to follow the head, which thinks only of escape.

Squinting further out on to the lake I see to my delight that the tufted ducks have given birth to twelve dark brown pom-poms: two per chick. In a few weeks they'll have become more stream-lined, but at the moment, similar to the moorhens, head and body seem only to have an accidental relationship, and the former is almost as big as the latter. This makes for entertaining viewing — in their excited haste to dabble and snap at gnats, the top-heavy ducklings frequently overbalance and end up face-down in the water, before their madly scrambling legs and natural buoyancy set them right again.

While her chicks are getting to know the water, the duck watches over them with the greatest vigilance. Her head and neck are in constant motion as she peers alternately at the sky and the shore; only rarely and for the briefest moment does she drop her own bill beneath the water to extract a piece of weed. Tufted ducks are usually silent, but when a gull glides over-head she makes a squeaky call of consternation, like the sound made by a saw when you try to waggle it loose it from a block of wood. The gulls are a real danger to her chicks, as confirmed by the signs fixed to the railings that urge you only to feed the ducks seed, rather than the bread that might attract more gulls. But what the duck knows and the council doesn't want to adver-tise is that after dusk the lake shore is alive with rats that come for the spilled seed but also wouldn't scruple to take a duckling.

If previous years are anything to go by, the mother will do well to raise even a couple of these chicks to adulthood.

I'd taken on some work earlier in the month, so it had been nearly a fortnight since my last visit when, towards the end of May, I found time one morning to resume my rounds of the local nests. I first visited the doves, but the mother was no longer sitting, and though I went further down the road and did hear the fey wail of an adult coming from behind the houses, I couldn't work out whether there was one bird or two, much less whether any chicks had fledged. Feeling unexpectedly desolate, I then hurried on to the great tit and coal tit nests in the park, but with the same result. I couldn't see signs of disturbance around either nest site: there were no feathers scattered outside the entrance to the rockface, and the nettles still stood upright before the crack in the alder. In any case, I wasn't too worried about the tits, which are known to be able to time their breeding to coincide with fine weather and the likely maximum supply of insects. I'd probably just missed the optimum week. But what about the doves? In my mind I saw again the eye of the sitting bird gazing back at me through the sight-line of leaves.

I never did find out what happened. I went back several times over the next few days, but saw no sign of young doves, so I assume the nest to have failed. Try as I might, I couldn't find the great tits either, but I had more luck with the coal tits. One day, while cycling past the lake shore, I heard a commotion in the trees, and when I stopped, and after a bit of waiting, the noise was revealed as the coal tit pair, pursued with flutters and incessant squeaks by four chicks. This was so close to their nest site that I think it can only have been the same birds. After watching for a while, I cycled on, meditating on the rewards and anxieties of parenthood.

Heron

7. Thunder at Solstice

It's late June, and tomorrow will be the solstice – a turning point, quite literally, as the earth passes closest to the sun and daylight stretches across eighteen hours of the twenty-four. The park is now basking in its own zenith; its slopes, ground by glaciers 10,000 years ago, are now a waist-high tide of grainy maroon grass. The whole hill-flank is furred with it – or rather them, because on closer inspection many different kinds of grass grow here. Some, sharp-stemmed and spiring, are still more or less green, but others are coloured silver and roasted ochre, and yet others, frail and diaphanous, look like nothing so much as purple smoke barely anchored on the thinnest of stalks. When a breeze pushes up from the sea it flares the grasses and makes the colours throb and fade, like bolts of colour pulsing through the body of a squid.

I'm walking these meadows an hour before dark. Vapour rises from the surrounding trees and slowly becomes visible as it thickens, a thin blue film motionless over the vale. The air is fragrant with pollen, ozone and plant oils: the smell of the planet. Walking through a meadow takes practice, and this is the fourth evening that I've come – nonetheless, every evening feels like starting again. I'm not a complete novice; when I was younger, I used to walk in meadows often. I go slowly and quietly, heel to toe – the 'stalker's walk' – attempting to move as fluently as the tussocky ground will allow. The mind, however, betrays the body, and a restless walker is a clumsy walker.

The first rule of walking through a meadow is to leave your mental clutter behind you. The more abstract knowledge you possess, the more clutter there is. I stood here a few evenings ago and heard this same bird. 'Chiffchaff' I said, naming it silently. I walked on, naming all the while: 'meadow brown', 'wren', 'alder', 'ringlet'. It's probably impossible to kick this naming habit entirely, but I can at least try to tone it down. This much feels necessary, because namers can't be be-ers. Which is to say, if you can stop identifying things then you have a better chance of identifying *with* them. In any case, my knowledge is entirely inadequate: there must be eight or ten different grasses here – perhaps more that I can't even see – but I don't know what they're called. I think about taking stalks of each type home and looking them up on one of those fancy apps that will instantly identify flowers for you. Human cleverness has no end, but awareness needs work, and isn't so easily come by.

This part of the park doesn't normally look like this, but the pandemic has cancelled the festivals that in other years leave the grass yellow, sweating and distressed where it's been trapped beneath stages, portaloo terraces and the thick white tripe of the marquees. The quietness of the park this year has also been good for birds. It's only my impression, but there seem to be more of everything – more tits, more nuthatches and more woodpeckers, like this green woodpecker fledgling questing for ants on the margin of the downslope. A funny-looking bird: a month old and already prehistoric.

I discovered the woodpeckers' nest hole late in the spring, just as the two young were preparing to leave. It was given away by the hysterical laughing call of the adults as they clung to the bosses and goitres of a nearby oak. Less strictly woodland birds than the great-spotted woodpeckers that also live in the park, the greens need plenty of wide-open grassland in which to grub for ants. One of their old names is 'rainbird', as their strange cry was supposed to portend a change in the weather. It's curious that the same foresight was also imputed to the mistle thrush – known as 'storm cock' – a bird quite unlike the green woodpecker in appearance, yet fond of the same habitats and from a distance often confused with it due to their near-identical flight patterns.

Mistle thrushes tend to be overlooked. Given that they're the largest of our resident thrushes, and quite noisy with it, I think this must be because they generally shun human habitation in favour of this kind of open country. The only place I've regularly seen them near buildings is in churchyards, where they come for the frosted red goblets of the yew trees, but they're just as fond of ivy berries, and as summer turns to autumn they can sometimes be seen as a pale flash as they manoeuvre themselves within the dark foliage. Mistle thrushes may be shy of people, but they're often very belligerent among other birds. Their size places them at the top of the songbird pecking order, and they'll use it to intimidate other species that they judge to have infringed their space. Their call when upset is a harsh, dry rattle, a bit reminiscent of the football rattles of long ago, and every bit as loud and tribal. They use it not only to chase away smaller birds, but also to defend their nests against the magpies and crows that are their chief enemies. This alarm call is one of the least appealing sounds that a British songbird can produce, but their song is surely one of the strangest and most beautiful. I say 'strange', because the tone of mistle thrush song is unique among our native birds for its evocation of all things high, grand and distant. This is hard to explain, and perhaps best attempted by contrasting mistle thrush song with that of its closest resident relative, the song thrush.

I fancy the difference to be similar to that between epic and lyric poetry, and I'm afraid that poetry of a kind is what I'm going to have to try to work up if I'm to do justice to the birds and what I mean. Song thrush song, especially at its wildest on the May dusks when the bird gives itself entirely to singing like one possessed, is all about self-expression. The range and clarity of its sounds is enthralling, and it's often claimed that only the nightingale can rival it for its emotive effect. The song is dark gold, but also pliant, as if it had some greenness in it, urgent, yet in its doublings-back and expectant pauses also seemingly seized by self-doubt. It's a dark confession, full of ecstasy and sobbing, played on a high-strung lyre.

Mistle thrush song is a different thing altogether. There's not a shred in it of anything personal. If the song thrush wishes to confess its own hopes, fears and agonies, the mistle thrush sings only in the strict metres laid down by tradition. There's nothing of darkness, gold or green about mistle thrush song; instead it's high and silver and somehow always far away. In fact it's a strange thing, but you can be standing right under the tree in which a mistle thrush is singing and still it sounds like it's coming from the other side of the world, or from the gates to the otherworld itself, or at any rate it's not addressed to you, insignificant mortal that you are. Song and mistle both clearly speak thrush, but the former scribbles it down in passionate, desperate handwriting, while the latter engraves its notes on the air in Times New Roman.

The traditional belief that the mistle thrush sings before the onset of a storm has now been vindicated by science. As their migratory records have shown, birds are sensitive to changes in atmospheric pressure; the lower the air pressure, the lower the density of air molecules, and the more effort it costs birds to fly. So the sight of normally very aerial birds such as swallows choosing to perch may indeed indicate an imminent change in the weather. But while less dense air discourages flight, it provides an additional incentive to sing, since sound waves travel through it more efficiently. This might help to account for the fact that some songbirds, such as mistle thrushes and blackbirds, seem actively to favour the atmospheric conditions immediately preceding and following rain.

It does feel like there might be a storm on the way, for a great mass of slaty cloud is building to the west. The wind that I first noticed tousling the grasses on the hill-brow is now gusting stronger, and smells of rain. The distant alders sway in slo-mo, and if they were closer I'd hear their thousands of leaves hissing like a freshet over stones. But isn't 'wind' a bland word? A dull, monotheist abstraction? Ancient people, who spent more time outdoors, knew that winds are plural. Supplementing their full-time winds of the cardinal points, the Greeks had an hourly paid pool of breezes – mostly women, of course – called the Aurae, whose remit was any more minor disturbance of air. These local winds can be very mysterious and unpredictable. Emily Dickinson's poem 'A Wind that rose/though not a Leaf/in any Forest stirred' is an accurate analogue of their eeriness, which I've experienced often, and which never fails to rouse some unknown faculty of my being, mothballed by modernity, into a state of high alert. It was very sensible of the Celts to ascribe these winds to hidden agencies, and in parts of Ireland a sudden sport of breeze is still called a Sidhe Gaoithe: a 'fairy wind'.

At this mid-point in the year, fewer birds are singing; only the species that commonly have second broods, such as great tits, wrens and blackbirds, need continue with territorial song. Most birds are absorbed with eating, insect numbers now being at their peak. But the first broods, though fledged, are still dependent on their parents for food, so I descend the hill to the wood to see whether I can find the great-tit family or any of the other birds that may still be foraging in family groups.

At the entrance to the wood is a stile shaded by a hawthorn. Flies shuttle to and from the sunlight, some pausing on the sun-warmed wood of the fence to rest and wipe their eyes. Perhaps because of the shelter provided by the hawthorn, the wind is more listless here. I'm reminded again of a deity, the Roman Cardea, who was a goddess of both wind and thresholds. Ovid tells how the virgin Cardea was raped by Janus, and in doubtful compensation given the right to preside over hinges. The pledge was sealed by a sprig of hawthorn, which thereafter became a tree much associated with gateways, portals and magical transitions generally. However remote we might feel from this story, it's true that hawthorn is often found at the entrances to woods, marking a place, moreover, where the wind is saddened and quiet.

In the half-shadow of the wood, the seasonal crux is more noticeable than it was on the hill. On the one hand, the great trees stand now in their summer magnificence. When bare of their leaves in winter I find it difficult to tell many species apart, but June grants them full distinction. Every tree is fully robed and at full stretch, the sap welling through trunk and bough having pushed leaves out beyond twig-tips to rustle and overlap in their own shade.

But even as they stretch out in fulfilment of summer, the trees are already preparing their autumn fruit, and since I'm no expert at identifying trees, I find these fruits the easiest way of telling the different kinds apart. All the trees in this part of the wood have fruit resembling exhibits at a museum of medieval life. I come first to a horse chestnut, displaying its soft-spined maces that will soon swell and harden into formidable containers for the glossy conkers of September. When the conkers are at their ripest the acorns will also be out, each one half-sunk in a brown-ish goblet, but in this early phase of their growth, nut and holder are still undifferentiated. Green, squat and ellipsoid, they look

like nothing so much as tiny knobbled quernstones. The sycamore seeds, meanwhile, twinned at the middle, are soft green
yokes for inch-high oxen.

Going deeper among the trees, I'm spotted by a jay, which
immediately alerts the whole wood to my presence with a barrage of horrible shrieks. How can such a beautiful bird make
such an abominable noise? In a competition to find the worst
singer among the common British birds there would be several
contenders. Crows are bad, gulls are worse, but for me the jay
would be the hands-down winner. And as in this case, beyond
the cacophony of the squawk itself there's the added irritation
that once you've been marked by a jay, everything knows that
you're there. Wear dull colours, avoid sudden movements, tread
silently, cat-like over the softest pine needles, but it's all a waste
of time if a jay has seen you coming.

For many years this awful screech was the only sound I heard
these birds make, so naturally I assumed that it was all they could
do. Then one day on the edge of a wood near where I used to
live, I heard the most delightful introspective chattering, melodious and preoccupied, as though an infant were babbling to itself.
I crept nearer and found the singer was a jay, for once oblivious
of my presence, gabbling on among the shadows of the lower
leaves. Since then, knowing what they're are capable of, I've
resented the jays' abrasive, incontinent alarm all the more.

On this occasion, though, the jay's vigilance works in my
favour, because its shrieks set off a domino effect of other alarm
calls that soon lead to something well worth seeing. First comes
the inevitable wren, chiding me invisibly from a rhododendron
tangle, then a series of blackbirds detonate along the path, one
female leaping up to my eye level the better to abuse me. A nuthatch joins in, but when I raise my binoculars to look at it, I see
that there's not just one bird but six. That's when I stop walking

and lean back against a tree, waiting for everything to calm down so that I can watch the nuthatch family at my leisure.

These birds are making good use of their special talent for walking down a tree trunk just as well as up it, a feat for which they're equipped with three strong toes pointing forwards and one backwards. Not even woodpeckers and treecreepers, whose claws are similarly equipped with grooves and strong muscles for gripping, can perform such a stunt. Once the nuthatches have stopped worrying about me, two fledglings, flat to the bark and facing up the trunk, begin to beg, and are soon fed a wodge of insects by one of the adults who faces them the other way; meanwhile the two other siblings have positioned themselves more conventionally on a horizontal branch and wait to receive their food from the parent foraging higher up the tree. I'm able to watch them for several minutes, listening all the while to the insistent squeaks of the young ones and the occasional strident piping call from the adults.

I emerge from the wood on the other side, opposite where a school playing field meets the park at a flaking metal fence. No matter what the season, the fence always seems to wear a mislaid hat or mitten on one of its ornate spears. All spring the playing field went unmown, and the grass and dandelions put the daisies in the shade, even hiding the parties of starlings that sometimes whirled down into it from their roost-cum-nursery in the derelict bowls pavilion. Now, however, some lockdown restrictions having been eased, mowers are at work, and two-thirds of the field lies shorn in strips of heaped stalks, with rough ovals around the tree trunks where the blades couldn't reach.

Even before the clatter of the mowers has faded over the brow of the slope, squadrons of jackdaws descend to waddle through the damp cuttings, where thousands of disoriented and pre-minced insects are the easiest of prey. In between stabs of their bills, they keep up a constant dialogue with their fellows on the ground and in the sycamores and lindens that border the field, a soft artillery of squeaky caws and agreeable explosions, rather like corn popping under the black lid of thundercloud that's gathering weight all the while. If the downpour comes, the jackdaws will retreat to the trees to wipe the wet grass strands from their bills and bide their time, because they know that afterwards the feast in the grass will be richer still, as earthworms rise to the drumming of the rain and flies of all kinds are loosed into the sodden air.

Myself, I've no intention of being caught in the storm, and I make for the park gate, but on my way out I can't resist taking a slight diversion past the lake. The bowling-green starlings have fledged their young now, and the youngsters, pale brown and slightly streaky, are making a terrible racket on the roof of the pavilion and in the lower branches of an alder that overhangs the water. Round one side of the alder a treecreeper edges silently and somehow nervously, like a caretaker on a late shift who's forgotten that the hall is in use for a school disco. The delinquent starlings keep up an incessant rasping as they beg for food, and sometimes they take short but perilous stiff-winged flights low over the lake to the central island, where scores more of them wheeze and rattle in the wind-tossed trees. The noise is quite something, and I can smell the birds, too – a faint stink of guano blending with the smells of lake-water and the coming rain.

I glance down into the water; the reflection of a gull high above me sculls just beneath the surface, then swoons weirdly in shallow loops as it's interrupted by ripples made by a passing moorhen. I know there are fish down there, because you get herons and cormorants on the island, and for a few weeks last summer a little grebe spent a week or two haunting the shallow elbow of weed-strewn water where people don't usually congregate. But the light reflected from the sky normally creates a dazzle that makes fish difficult to see. Only now, with the lake shadowed by the looming rainclouds, can I make out shoals of dark, blunt-headed shapes darting about. I'm still watching them when the clouds finally break, and I'm sent hurrying for the gate.

The next morning dawns bright, and birdsong rings clear in the rinsed air. Through the drawn curtains and the background babble of the radio I hear the wood pigeon earnestly doing his breathing exercises, and the soft, ashy caws of jackdaws mediated by the brickwork. Shinier in the foreground, a blue tit gives its cross, silvered churr from the apple tree whose leaves are lacquered with summer rain, and I catch myself smiling when the sound gives me a picture of him restlessly rotating his blue-capped head like a tiny, caterpillar-powered R2-D2. The news comes over the radio that Leicester – a city of almost half a million people – will have to go back into lockdown, after suffering several new outbreaks of Covid-19. It looks increasingly like recovery will be a long haul.

As I half-listen to this, a male blackbird lands on our back wall with a flick and a flourish, pursued a second later by a brown, half-speckled youngster. The chick, perhaps two months old, makes an odd, reptilian whistle and then bounds along the top of the wall in a low posture of supplication. It's only now that I notice the worm-segment drooping from the bill of the adult. Turning to the chick, he proffers the worm, and the young bird angles its head to accept, gaping a wide, orange-yellow chevron. I wonder that there is only one chick, but perhaps there are others with the female, wherever she might be.

Both here and in the park it seems to me that the end of the breeding season has allowed a new mellowness to creep into the birdsong. The park blackbirds extemporize freely, apparently more for their own pleasure than out of any competitive or possessive urge, and even the beechwood wren sounds a little less tart. I'd swear that its song has slowed down, as if the effort of contracting muscles, the stopping and unstopping of the syrinx and the gaping of the bill were altogether a bit too much trouble.

Amid this relaxation only the song thrush seems to be going the other way, raising its song to a new intensity even as the others slacken off. A month ago, I saw the thrushes far more often than I heard them, as they came shyly from the shade to take worms from areas of grass the more aggressive blackbirds had left vacant. Now it's a different matter, and theirs is a constant voice in the dawn chorus, throughout the day and into the dusk, with the force and variety of their song seeming to build all the while – the matinee only a pale rehearsal for the twilight show-stopper. This evening I'll return to the park in the afterglow of the departed sun and hang around the lake until bat-light. I'll be there when the finches stop singing and the wren retires, and I'll keep still while the blackbirds have their mad half-hour on the margins of the roost. I'll watch the first bats beat out from the trees and listen while the robins tick, pause and tick again, as though searching for the combination that will finally open the vault of the night. But above all I'll come for the weird, unsettled ecstasy of the thrush.

I'm suddenly lifted from these thoughts and returned to the present by a bird-call that I hadn't expected. It's a rising two-note cry, and it takes me a second to place it as a curlew. In winter this is a bird of the shore, and in summer it shouldn't be here at all, but on its high moorland breeding grounds far inland. So what's it doing here now? My perplexity only lasts a second, because the call comes again and I realize that it's only quite a poor impression of a curlew – to be more precise, it's a curlew call thickly accented with blackbird.

Several of our commonest birds are able to mimic other sounds, and, as I've said, in the absence of any plausible evolutionary advantage, it has to be concluded that they do so because it appeals to them. This is supported by recent research suggesting that mimicry was unknown to the ancestors of our songbirds, who must therefore have picked up the skill somewhere along the evolutionary road from archaeopteryx to ventriloquizing blackbird. The most likely sounding of the scientific hypotheses for mimicry is that female songbirds confer favours on the most inventive singers, but the logical conclusion for this line of reasoning would be no future for species-specific songs at all, and instead a sort of Esperanto of more or less improvising birds.

That the mimicry is basically playful rather than practical is hinted at by the random way in which the copied calls are integrated within the singing bird's usual repertoire. While this blackbird is plainly imitating a curlew, he's just as clearly not trying to pass himself off as one; if this were a serious attempt at identity theft then he wouldn't have followed his curlew impression with a tell-tale sequence of notes that could only have come from a bird of his own species. Indeed, it seems unlikely that the blackbird knew that his borrowed call came from a curlew at all. The rather indiscriminate nature of mimicry supports the idea that the birds just tune in to whatever ambient sounds take their fancy.

If, as it's proposed, this talent for mimicry has been built up over thousands of generations, then one wonders what starlings sounded like in the garden (or car-park) of Eden. Was this a world of sonic purity – a sort of avian nativism or reversal of the myth of the Tower of Babel – in which each bird stuck rigidly to its own tunes? The only explanation I can really believe is that the ability to mimic is a by-product of the increasing complexity of birdsong, which may well have been grounded in the need for differentiation but finds no sensible or necessary culmination in a twenty-first century starling sounding like an ambulance siren. Evolution continues all around us, inexorable as a glacier, but bearing along within it an unpredictable freight of deviance, randomness and whimsy.

I spend the late afternoon with the children, inventing games to keep them away from the computer screen. Then there's some shopping to do, which still involves queuing outside the supermarket at two metres distance from everyone else, and, once inside, following a one-way system up and down the aisles, trying not to be too conspicuous about holding my breath when passing close to other shoppers. (At this stage of the summer, it's still not compulsory to wear a mask). After the bedtime routine I take up my binoculars and head back down to the park.

Even in a smallish urban park there's something beautiful and unnerving about the dusk, though that one word, evocative as it is, can't hope to convey all the estuarine subtleties of tone as the light seeps slowly from the sky. Shadows seem to well up from the earth itself and flow out from the bushes and thickets where they've waited patiently all day, caged by light, now to blend themselves and run over the grass and footpaths, where beneath the moon the gravel takes on an answering tinge of lunar white. I have the impression from novels that British people used to go outdoors at twilight much more than they do now – just as they still do, perhaps, in Mediterranean countries. If that's so, then we're missing out, because the beauty and mystery of this time of day – less a time than a *state* – has a peculiar effect on the soul, both soothing and gently estranging, that seems a balm against the perpetual haste and heedlessness of our daily lives in a consumer society. Both we and the world put on strange new costumes at dusk.

Around here, and just at this particular moment of summer, the dusks are muggy and green, heavily drugged with the odours of hawthorn bloom and early lilac that hang so thickly about the flowers as to exceed smell altogether and briefly stimulate some neglected and unnamed sense halfway between scent and sight. Some people pass along the path, but they don't see me in the semi-darkness of the trees. Headlights swing around the bend, and shadows pounce out sideways, then relax. The blueness has all but drained from the sky, and the portion that remains, jagged-edged between clouds, is already grimy behind thickening motes of dusk, like a soiled sherd of Wedgwood dug out of a midden. A brief commotion on the lake might mean that a rat has swum out to where the moorhens are roosting on the island. But then, at last, all is silent, and the thrush has the stage to itself. There's no hope of my seeing it now, but that doesn't matter in the slightest, since its appearance is far exceeded in beauty by the envoy of its song. I listen to its weird, unsettled ecstasy and inhale the mingled smell of trees and dank water, until the midges get the better of me and I make for home and bed.

Though by the age of nine or ten what I'd been told about the solstice entered my official knowledge, like most children I also inhabited an alternative version of time whose cyclical and, as I imagined, eternal course was marked by events in the natural calendar. Within this parallel track of time, at once vernacular and cosmic, the true zenith of summer was signalled by flying ant day. Our cul-de-sac was built over fine, sandy soil of a red-brown colour, and each June little heaps of it would begin to appear between cracks in the paving stones and especially around the thick metal slabs that covered the water stop-cocks. These tiny workings revealed the labour of the ants that, out of sight of us surface-dwellers, were preparing for the most important event in their own calendar.

The ants were of two kinds: the common, medium-sized blacks and the scarcer, smaller reds, which now I come to picture them weren't really red at all, but more a sort of amber colour, quite dark when seen against the ground, but translucent if raised on a finger. I'm afraid that we children weren't always kind to these animals, and weren't above dousing them with water, rarely but inexcusably boiling. Not only that, but we sometimes exercised our fascination by carrying a couple of reds from their nest and dropping them among the blacks, or vice versa, to watch the ensuing fights. Thus were the cruelties of the Colosseum replayed in miniature in a Birmingham suburb. Although the reds were smaller, we soon found out that they were fierce beyond their size and willing to bite, and we were nervous of handling them in a way that we weren't with the blacks.

Throughout June, whenever it was too hot for football, we would prise open the stop-cock lids or sections of damaged flagstone to view the ant city growing beneath, and after a while we noticed that they were protecting tiny, soft, off-white eggs. We were rather repulsed by this, and quickly lowered the slab, but still it was exciting, because as the number of eggs grew and the displaced soil overflowed its original small cones to spread mini landslides into every crack, we knew that it was nearly time for the ants to fly.

The way I remember it now, there were three final portents. First was the emergence of much larger, beetle-sized soldier ants from somewhere underground. (These forbidding-looking insects became forever linked in my mind with the black-armoured imperial pilots from the first *Star Wars* film, only a decade old, which I'd just been allowed to watch.) The second signal was the weather, which began to grow unbearably humid the day before the ants would fly. Sweaty and listless, we gave up all efforts to play and sat down on a neighbour's wall, keeping one eye on the cracks between the paving stones.

But we knew that the waiting was over when the birds began to gather. The swifts had been with us for two months already, and by now their numbers had been doubled by their fledglings. Small parties of adults and young would dash back and forth in tight, screaming knots, playing chase in the clouds or lower down between the chimneys. Unlike the swifts, starlings, jackdaws and black-headed gulls were usually rare above our road, but now we became aware of them perched on the aerials or, in the case of the gulls, gyring round gently in the upper sky where heavy black clouds built along the horizon.

Then, finally, it happened. Magically, as it seemed, winged ants began to pour out of the earth, some with their wings neatly folded on their backs, others tottering under wings prematurely raised, so that the insects nearly overbalanced altogether, as if a tiny boat had hoisted a sail twice as large as itself. And as the pavement leaked ants, so the birds' banquet began. The ants' woes began even before take-off, as sparrows bounced among them, pecking freely and not even pausing to chirp their pleasure. For such squat-postured birds sparrows can move with surprisingly dexterity when they've a mind to, and while some were content to collect ants from the hot flags, others performed little leaps and twists to snap them up in mid-air.

But the real feast took place higher up, and I remember being impressed at the orderly way in which the different species shared the ants without conflict. At lowest altitude came the swifts, squealing with glee as they opened wide their whiskered gapes to take in the ants at 40 miles per hour. Above the swifts, the starlings and jackdaws shared roughly the same airspace, while at the highest level of the ants' ascent they were picked off by the gulls, which applied to their feeding the same unexpected deftness displayed by the sparrows. In my memory, this feeding frenzy always culminated in the breaking of the storm, and from then on the character of the summer changed; though our daily activities remained much the same, we were now on the countdown to autumn.

It can't be long now before the ants start to fly here too. When I notice them gearing up, I plan to bring the children out to see them. We try to give them an outdoor life, but it isn't the same as when I was young, when my friends and I virtually lived on the street and only dashed home for refuelling with colossal mounds of pasta. But still, I think they'll like the ants – who knows, the sight might remain with them as it has with me, and give them something other than the pandemic to remember the year by.

Chaffinch

8. Birdsong in the Blood

In the eighth century, half a millennium before the anonymous author of 'Sumer is icumen in' wrote his verses, the great painter and poet Wang Wei was sent away from the dynastic court on a mission to China's troubled northern frontier. He was alone and miserable in an unknown land, and struggled to understand the strange dialect of the locals, but while journeying one day he heard a bird – an oriole – sing the same melody that they sang in his own part of the country, 'familiar as a friend'. He may have been far from home, but the oriole's song, heard since childhood, let him know that he was not entirely outcast. And since the poem in which he recorded his feelings has survived, Wang Wei's oriole is still in a sense alive to be heard by us today.

Birdsong is not only natural, but also cultural. I don't just mean that birdsong is a theme for culture, although of course that is true as well: since Wang Wei's time and also long before, it has inspired countless poems, songs, tales and compositions in music. A book could be written solely about the utterly different ways in which the English Romantic poets John Keats and John Clare wrote about the nightingale. But birdsong is cultural in other senses too. For one thing, because at least some species display sonic preferences, are able to improvise and can also mimic notes and phrases that they find appealing, they don't just make predictable sounds at the behest of their genes but also have what we're obliged to call an oral tradition.

A few years ago, researchers in New Zealand who were studying the songs of the yellowhammer, a beautiful little finch-like bird introduced from England two centuries earlier, discovered that their calls were markedly different to those made by birds still resident in England. The main yellowhammer song is thought to lend itself especially well to fair reproduction in human (English) speech, and in many bird books you will find it set down as 'a little bit of bread and no cheese', which if you extend the cheeeeeese in a beseeching way does convey something of the song's rhythm and tone. But like those of other species, yellowhammer songs also possess variations or 'dialects' that generally go unremarked by the casual listener.

Although the New Zealand birds were conversing in fluent yellowhammer, it came to light that their repertoire comprised a greater number of dialects than are found in yellowhammers still living at the sites from which the birds were originally captured and sent to New Zealand in the nineteenth century. There seemed only two possible explanations: either the Antipodean birds, in the course of their long residence, had picked up the local twang, or else they had remembered and continued songs that the Englanders had since forgotten. As yellowhammers seem not to take up new tunes readily, and because continental European yellowhammers still sing in dialects similar to those practised by the New Zealanders', the researchers concluded that former English yellowhammer dialects were indeed still being preserved on the other side of the world.

In one way, perhaps, we shouldn't be surprised by this. After all, the same phenomenon is noted in human languages – for instance, the persistence of rare or obsolete Old World accents and dialect words in Newfoundland. But while we take oral tradition and transmission for granted in our own species, we're less accustomed to granting the capacity to others. And yet here we have it: a scientifically supported example of birds passing on dialects to each other – in short, culture. Why the English yellow-hammers have a diminished range of songs is something that it's hard to be sure of, but like the rest of our farmland birds their numbers have declined catastrophically since the industrialization of agriculture, so it seems likely that the relative paucity of their dialects simply reflects a much smaller and more fragmented population. If that's a melancholy fact, then consider the species skilled in mimicry that still include within their repertoires the songs of other kinds already lost to extinction. Several such cases are known, and there seems no reason to doubt that in many other instances calls now regarded as characteristic of a particular species originally belonged to a quite different one.

So birdsong isn't only passed on among individuals of the same species, but is also transmitted across what we're accustomed to thinking of as fixed species barriers. Once we accept the idea of birds as cultural beings, this opens the question of how their cultures overlap with our own. Human beings have been listening to birdsong for a very long time, long before almost anyone would claim that there were human beings at all. Recent studies on fragments of bird bone show that the first known songbirds evolved in Australia over fifty million years ago, while at the most generous estimate the ancestors of human beings have only been around for five or six million. In a sign of the power still possessed by science to either repress or validate more vernacular forms of awareness, the announcement of this discovery quickly prompted retrospective exclamations to the effect that we should have guessed it all along, since many Australian songbirds display an intelligence that must have taken a long time to develop. This intelligence assumes many forms, from symbiotic relationships with other creatures to the use of tools, but is most evident in the complexity of Australian birdsong. This complexity is demonstrated across the range of songbird species and also finds its individual champion in the lyrebird, a creature gifted with such an extraordinary range of sounds that an Aboriginal myth credits it with the ability to understand and communicate with every other species.

Since our awareness of birdsong stretches far back into prehistory, we can evoke a third category of memory that both contains and underwrites what we think of as individual and cultural memories – a category that we might call species memory. Tens of thousands of years ago, when people cohabited on the British Isles with many species no longer present, including another species of human being, it's thought that we may have wintered in caves by the seashore and moved inland again in search of food each spring. Cave-born children of the winter months would have heard the cries of seabirds even from the womb, so what reason is there to doubt that hearing them again each year as an adult would have brought sensations of belonging and comfort akin to those recorded by Wang Wei on hearing his oriole?

There are many other ways in which birdsong has influenced human cultures. During the last Ice Age many more geese and swans than now would have migrated to southern Britain from lands further north. Did the peregrinations of our ancestors follow those of the geese, and did the strange clamour of these birds, particularly eerie after dark, inspire the strange stories they told each other on their long journeys to the sea? I don't think that any of this is too fanciful, and the apparently unconnected appearance in different European folk traditions of spectral, night-flying hounds might be submitted as suggestive evidence of the effect of bird calls upon the human cultural imagination.

So it's easy to see that our millennia-old exposure to bird calls and songs must have influenced us in myriad ways and on different levels of our awareness, both individual and social. The discovery of the Australian origin of songbirds confirms that the intelligence of our own species developed in the context of the pre-existing intelligence of birds, and there is every reason to believe that we learned much from them in matters of finding food, avoiding common predators and also, of course, singing – or at least making conscious, meaningful sounds. Listening to birds would have helped keep early humans alive, just as it still does for other species, but there must also have come a point when birdsong was valued aesthetically, as a source of pleasure or comfort in its own right, and capable of restoring meaning to human lives shadowed by sadness.

This isn't just speculation. Clinical studies have shown that listening to birdsong reduces stress, increases attention capacity and benefits mental health generally. Of course, not every sound made by birds is received by us as soothing, and it would be fascinating to know whether our general antipathy to birds' alarm calls is solely a response to their sonic harshness or perhaps also awakens a memory in our bodies of a time, millennia ago, when noticing and reacting to such calls might have been of vital interest to our own survival. However that may be, the increasing evidence that at least some kinds of birdsong are good for us surely only gives specialist endorsement to what we already knew. During this spring and summer, countless testimonies online and in print have expressed the peace and enjoyment that people have found in birdsong, and its subtle power to ameliorate the anxiety and trauma caused by the Covid-19 pandemic. If we ask why this should be, then the most straightforward explanation is that we've grown up with birdsong, both individually and as a species. It has always been there, and it's part of our feeling of belonging to the world. And since sounds produce chemical effects within our bodies of stress or pleasure, it's more than figuratively true to say that we have birdsong in the blood.

*

Thirty-five years ago, my walk to primary school took me beneath an aqueduct that held for me both apprehension and deep fascination. There must have been several leaks in the structure, because viscous, blackly gleaming water dripped down even in the hottest of summers. In company I used to play at dodging the drips while my friends and I hooted inanities as loudly as we could to force the rather disappointing echo. Alone, I feared the place a little, partly because of the chance that the drips presaged an imminent collapse, but mainly I think because of the strange mood of the place, a tiny microclimate of gloom, traversed in ten footsteps but sinister in its perpetual shade and dampness, and the streaks of moss and slime that stained the rough sandstone blocks. Although I knew this moisture came from the canal that lay coffined in its narrow channel above, I fancied that it oozed from the walls themselves, as if the stones were sweating thick, dark water.

One day I went to press my palm against the stones when I noticed a sound coming from a crack a little further up. Standing on tiptoe I could just make out a number of baby birds, now silent, yellow-lined frog-gapes clamped and sagging, as though someone had removed their dentures. It was a wren's nest, hungering in the wall. I was enchanted; I'd never seen anything like this before. But now I came to think of it, I'd often heard a wren singing from just this place – from the canal embankment or from the heavily littered hawthorn bushes on the flank of the slope. I remember retreating to the opposite wall and keeping as still as the nestlings until I was rewarded by the song of a parent bird resounding within the man-made cave of the aqueduct, and then the sight of it whizzing into the shade like a tiny brown shuttlecock to feed the young.

Perhaps it was shortly before the schools broke up for the summer, or maybe I avoided the place purposely for fear of upsetting the parent birds, but for whatever reason I don't remember what happened to the nest, so I can't say whether the little wrens fledged successfully. But it's a pleasant thought that their descendants are still singing under the same bridge, rowdy little pocket-trolls. I summon this memory with ease because it lies inside me, never far from the surface, to be stirred every time I see or hear wrens now. I encounter them almost every day, but each new meeting adds to, without ever replacing, that foundational experience of finding them under an aqueduct, far away now in space and time.

Another childhood bird memory brings back my grandmother's house as it was in the late 1980s. It was a semi-detached house in a large village, at that time filling up with new housing estates, but still fairly rural. House martins returned there annually from Africa to daub their nests using mud we watched them gather by the beak-full from a trampled strip of field behind the back garden, which led to a cattle farm. The birds twittered constantly, and if it wasn't from sheer neighbourliness then I don't know what else would have induced them to spend valuable energy in this way, for they worked from before dawn until after dusk. They were always up before me, and even in those days, when I could bridge the gap from sleep to full consciousness in seconds, and a lie-in was a chore, I would sometimes stay in bed for a few minutes to listen to their calls, which reminded me of the working of knitting needles. House martins are like sparrows in that their calls sound very much alike in concert, or at first listen, but reward greater attention with a surprising depth of variety. The individual martins produced twitchy, insect-like calls, somehow a little wet, as though they came from the cheeks, but they differed a great deal in duration and form. Most seemed to comprise two short rough sounds of slightly different pitches, like the *chiʒ-ʒick* of a wagtail, but others were a little longer and, again, what

I can only describe as *wetter*, like a microsection of a splutter. They made these calls either singly or in rapid little volleys, and there was also a great deal of variation in their tone, which ranged from conversational to urgent and alarmed.

A few years after the martins stopped coming, never to return, a third memory is of another African migrant: the nightjar. These singular birds, hawk-like in flight, but at rest resembling nothing so much as chunks of rotten wood, migrate here each summer but are rarely seen or heard, being strictly nocturnal. Around midsummer's eve, provided the weather was fine, my father would drive me, with some of his friends, to a vast undulating heath an hour's drive from home. It would be almost midnight when we got there, and when we took one of the tracks that led off around the pine brakes on to the ferny hills, we entered a weird world of half-light and silence. We'd see the looping display flights of woodcock in clearings of the forest, glimpsing them briefly by the light of the constellations as they flew in and out of the darkness of the trees. And it was then, if we were lucky, that the nightjars would begin to sing their weird, one-note reel, much more like a frog or a giant insect than a bird. We'd approach the sound as softly as we could, but if we came too close the sound would wink out like a suddenly extinguished light. Sometimes they would hawk across the heath in front of us in stiff-winged silhouette, their flight oddly buoyant, as though they moved through water – and indeed it really was like being on the sea-bed, down there in the pitch-black billows of the bracken, looking up at the birds that swam above us through the grainy gloom.

These three memories – of wren, martin and nightjar – are only the closest to hand of hundreds of recollections of birds and their songs. These memories begin consciously from the age of about three or four, when I watched those vast starling flocks

sculpting the greasy dusk above Birmingham before they set-
tled to roost. No doubt, though, I was aware of birdsong from
even earlier days, in a cot now lost in time or even in utero – so
loudly did the wood pigeons croon from our TV aerial. So, in
a modest way, these birdsongs have helped form my identity.
And if your memory is as bad as mine, you'll understand why
I also value them for the fragments of context, incidental to the
birds themselves, that they also save from oblivion: the sweating sandstone blocks in the aqueduct wall, the particular shade
of blue paint – not seen since – on my grandparents' outhouse
door, and the white-robed party of druids or wannabe Klansmen
we encountered one night while nightjar-hunting in the woods,
and who seemed just as alarmed at the sight of a party of birders
as we were by them.

And because these three bird species, despite their diminished numbers, are still with us, the memories are more than just
museum pieces of the individual consciousness but also constitute parts of a living awareness of the world that can be enjoyed,
recognized and shared with others now. If I see or hear wrens,
martins or nightjars, I'm unfailingly delighted, but part of this
delight is also invested in my own personal history and the shared
cultural history that I'm part of – the culture of, among other
surprising things, nineteenth-century aqueduct design, the damp
outhouse with its odd implements and huge rusted nails that
could have come from Golgotha itself, and the persistence of
paganism (or worse) in the West Midlands woods.

So the connection of birdsong to childhood is not at all incidental, and this is something else that tends to emerge from
studies on the beneficial effect of birdsong for human health.
Interview respondents say, without being prompted, that exposure to birdsong calms them down and restores their concentration
because it takes them back to the relatively carefree days of their

childhood, when they had the leisure and interest to notice birds and their calls. When I first read transcripts of these interviews, it felt like another of those discoveries that confirm something already long intuited, and I've no doubt that my adult fondness for the twitchy conversation of house martins owes something to a half-buried memory of listening to them from my bed in my grandparents' house many years ago.

I've already mentioned that some birds possess an uncanny talent for mimicry, with repertoires ranging from rival species to ringtones. Most amusing of all, perhaps, are those clever birds that copy human speech – crows with regional accents, loquacious parrots and foul-mouthed budgies. We return the compliment, too, with our own efforts to reproduce birdsong using our rather different vocal equipment. The importance of mimesis for our own species runs deep, and some scientists even believe that both human language and human music have their roots in our attempts to copy the sounds of the natural world around us. Whatever the truth of that, almost virtuosic imitations of birdsong are known from traditional communities as far apart as Tasmania and the Outer Hebrides; but on a more lowly level, who can honestly say they've never at least been tempted to huff like a wood pigeon or take off the droll, drooping whistle of a starling?

Or let's consider the magpie. Can you imitate one? Since all mimicry begins with listening, we should first look up a magpie's call on the internet, or better still, go outside and find a real specimen. But once we've done this, we still need to work out how to reproduce the sound. Throughout the stages of listening, reproducing, listening again and revising, it seems to me that the main challenge is of attention and accuracy, by which I mean giving one's fullest attention to the birdsong and then using one's utmost powers of expression to render this song truly enough for others who have heard it to nod or smile with recognition. In other words, the technical challenge serves a purpose that's ultimately artistic. And though it might well be of interest to record a song and then with the aid of a computer slow it to half-speed to reveal all its hidden slurs and glissandos, this can never be fully satisfying because in the end it's how *we* hear the birds that most counts at a human level. This returns us to the challenge of how to communicate to other people the sounds made by birds. The inadequacy of 'tweet' and 'cheep' is evident to anyone who's properly listened to them, and even onomatopoeic transcriptions are prone to national or individual variation, as with the case of the chiffchaff/zilpzalp/siff saff. But since such differences are inevitable, perhaps this problem is something of a red herring.

To return to our example, what sound does a magpie make, after all? Does it chatter? Not really. So does it rattle? Sort of – but to me it sounds thicker and more moist than a rattle. A shaken seed-head rattles, but a magpie's noise seems to have a bit of saliva in it. (Do birds salivate? I don't believe so.) It's perhaps more like a splutter, but that seems to go too far the other way, and splutters are usually involuntary anyway, which doesn't seem right. So perhaps the best we can do is coin a new verb, midway between a splutter and a rattle. Call it a spluttle.

So then how does one spluttle like a magpie? This is where things can get painful. First suck in your cheeks so that the inside of each can be clamped steady by your molars. Your nerve endings will tell you how hard to bite down. Now blow air from the back of your mouth, behind your clamped teeth, in order to vibrate your cheeks and you'll have a spluttle (or a chattle, or whatever it most sounds like to you). I'd recommend practising on your own before testing your imitation on someone else, because a misaligned spluttle can get messy, and there's a pandemic on. Best of all, test it on a magpie; most likely it will eye you with disdain and fly away, but what joyous validation it would be if it spluttled back at you. I've had this happen on occasion, not only with magpies but quite frequently with ravens, and before my voice broke I was a respectable enough buzzard that once or twice they condescended to wheel around and pass the time with me.

But fun and rewarding though it can be, human mimesis of bird calls quickly meets its limits, for even the most golden-throated of imitators can't ultimately overcome their lack of a syrinx. Our most successful attempts at oral pastiche must begin by modestly selecting the calls that are most reproducible with larynx, tongue and teeth. Sticking with magpies, although the spluttle is the bird's most familiar cry, it's by no means their only one. Being crows, magpies are full of surprises, and their range is a lot more impressive than a cursory listen would credit, incorporating a series of rasping, grinding sounds, less aggressive and sometimes quite amiable and conversational. These secondary noises are quite beyond my imitative abilities, so to translate them with any degree of fidelity I need to use a different cultural tool, one unique to our own species.

As you'll have noticed, I often resort to similes and metaphors to try to give as accurate an impression as I can of what a particular species sounds like to me. These word-images can be of many different kinds, but it's often struck me when listening to birds how frequently I receive them by analogy with objects and practices that are just as endangered by our uncritical acceptance of technology as the birds themselves. In other words, the similes that can help bring birds nearer to us depend on correspondences with sympathetic objects that are themselves becoming rare.

To use the magpie again, what I've just referred to as its 'secondary' calls are suggestive to me of sounds that I speculatively

associate with a cutlery manufacturer, or atelier of knife grind-
ers. And, shifting my attention to its visual appearance, I also
want to say that the way it raises and lowers its tail is reminiscent
of a pump-handle. But while these images might still just about
work as evocations of a magpie, how many of us have actually
heard the goings-on of a knife workshop, or witnessed water
being drawn up by a pump? Or again, doesn't a magpie also
look a bit like the clapperboard used in old movies? The director
shouts 'Take one!' (for sorrow) and a magpie brisks down on to
your garage roof. 'Take two!' (for joy) and the magpie flounces
and flicks that pump-handle tail. But the success of figurative
language depends on the existence of objects and experiences
shared between communicator and receiver – and nowadays if
we ever see a pump at all it will likely be in a museum, while
those movie clapper boards are only found in old Tom and Jerry
cartoons.

The same problem occurs when we think of what other birds
sound like: much of the imagery is either medieval or industrial.
A wren's song summons the whizzing of a spindle, a bullfinch's
call is like the soft creaking of the iron gate to the orchard,
migrating geese snicker and bray like coupled train carriages
going over points, and just two reed warblers can make a reed-
bed sound like a smithy. But all the sympathetic objects in this
list are rapidly fading from the world, if they haven't already
done so. Also, they're not being replaced by equally resonant
equivalents.

I think that an age of enforced digitization impoverishes the
common soundscape and also restricts our ability to communicate what birds sound like. It's only necessary to visit countries
where this process hasn't yet taken over to be reminded of the
sounds we've lost, from the muffled thud of the stamp that vali-
dates tram tickets (here replaced by the beep of a tapped bank

card) to the swish of park-keepers' brooms (replaced by the terrible racket of electric leaf-blowers, reminiscent of nothing at all in the natural world). Thankfully, there are still exceptions: name any method of opening an alcoholic drink and the sound can be matched to a starling, from a drawn cork to the click of a screw-cap to the snap and hiss of a ring-pull. Nevertheless, it seems clear that nature and a properly human culture are steadily going down together, along with the language that sustains and unites them.

Right at the end of July it's reported that the sudden decline in human activity during the pandemic has been registered by seismologists as a wave of silence passing over the earth, its course exactly following that of the virus. From China to Iran to Italy, vibrations from traffic, industry and construction work faded or for a time halted altogether; the crust of the planet ceased to judder with the noise that had been dinning, seemingly unstoppably, since the onset of the Industrial Revolution. Finally, the earth could hear itself think, and the voice of its thought was song.

During these months it's often occurred to me that there are two kinds of silence. There's the silence of shame, cowardice, inaction, loss and death; and there's the silence of contentment, rest and peace, which is also the sort of silence on which the attention can feed, and rediscover things it thought it didn't know. In 1962, Rachel Carson's book *Silent Spring* warned of a world whose birds were being poisoned by pesticides, and consequently a world devoid of natural song – the silence of death. And in large part this dystopia really has come to pass. In Britain alone, there are 50 million fewer birds than there were when Carson wrote her book; imagine the sound of 50 million songbirds serenading the spring and you get an idea of what we are missing. The thoughtlessness, greed and violence suggested by

Brueghel's nest robbers, and the 'murd'ring gang' that appalled Robert Tannahill, have expanded far beyond the individual and become ingrained in the way that we collectively treat the natural world. But the effect is the same as that described by Tannahill: in robbing birds of what they need to thrive we are also robbing ourselves of what John Clare in 'The Nightingale's Nest' called 'the old woodland's legacy of song'.

Humans have short lifespans and shorter memories, and an unhelpful tendency to think that how things are now is the only way they can be – that the present is inevitable and the future will look after itself. It's in this context that the wake of silence which followed the Covid-19 pandemic can be seen, despite everything, as an opportunity, perhaps even a gift, in the sense that it might restore our awareness of the natural world that is our only home. But this will only be true if the new attentiveness to what is really important in our lives can be retained and then acted upon, if and when the virus is brought under control and the profit-and-distraction machine begins to operate again at full power.

Robin

9. A Light in the Darkness

The first thing you notice is the mess. A winter wood resembles nothing so much as a discount haberdasher's, an end-of-season sale of offcuts and tired-looking forms. Bare of their leaves, the trees look like battered lamps that have lost their shades. Beech trunks seem swollen and grotesque, disreputably rain-stained at the groin. Poplars stick out of the mud like greasy feathers, and only the birches are almost as beautiful as they were last month. Naked without their smoky golden haloes, they bend their slim white branches into cages for the wind.

Autumn's celebrated 'riot' of colours has died away, and what remains is more like a peaceful demonstration of what can be done with an infinite variety of browns and a little winter sunlight. The basic *imprimatura* is dun, but it's hatched with darker tones, and even the weak white sun smoking somewhere overhead occasionally sends down enough light to make a lattice of shadows on the cheerful disarray of twigs, thorns, yellowed weeds and other winter sweepings.

At this time of the year most birds are lying low, and with so little cover, whatever you may see or hear has certainly seen or heard you first. Wood pigeons detonate from the side of the path with a colossal racket, tails skewed almost at right angles to their plump bodies, like axe-heads careering loose from their hafts. By the time a wren appears from nowhere and begins to chide and trill, my cover is well and truly blown.

Birdsong in winter is a very different affair from the massed orchestra of spring. With the breeding season over, the boundaries of territories are less ardently defended, and in many cases have disappeared altogether. The search for food preoccupies all species, and cooperation is more advantageous than competition. It's in winter woods such as this that mutual benefit societies of small birds can best be observed. I'm nearly at the corner of the wood when I hear the first outriders of a flock, approaching from somewhere deeper within. I keep quite still as they approach. It's rather like waiting for a wave to break; first comes the *tzee zee* of long-tailed tits, and then I'm overwhelmed in a bustle of movement and sound. Suddenly the whole wood, that had been almost silent, is fizzing with life. It seems frantic, but is actually quite orderly. Blue, great and long-tailed tits tilt, flit and pivot among the branches in search of insects and their eggs, calling all the while; nuthatches fuss around the bole of a fallen tree, climbing up and down head-first; blackbirds, their bills dull for winter, make sudden, explosive forays into the leaf-litter and toss aside the leaves in that slightly deranged way they have.

Among all this activity I stand quite still, having roughly the same relation to the birds as a maypole has to the children who dance around it on ribbons of colour and sound. This is the epiphany of the winter wood, so familiar from my childhood, and while I'm caught up in it I forget myself almost entirely. After a few minutes the effervescent wheel rolls off into another part of the wood, and I'm left alone once more in silence.

The change to winter, long anticipated, still seemed sudden when it came. In the robins' phrases there is that familiar and unfailingly poignant mix of cheer and sadness. Franz Kafka, when asked whether there was any hope, is reported to have said, 'Yes, but not for us,' and this seems to be the gist of the robins' clipped, bittersweet bulletins. It's difficult to place when we first heard that introverted note, and it seems possible that it was there all along, or at least from the day before the solstice, almost six months ago now, when walking in the park woods I first sensed that summer was on the wane. So the seasons incubate each other: beneath this wet, undistinguished humus, rotted to brown from the gold and vermillion leaves of autumn, next spring's bulbs are already charging in secrecy.

The virus that came with the spring has not gone away, and the whole country remains in some form of lockdown. Where we are, schools have had to close a week early before the holiday, and some shops and restaurants have shut for good, unable to withstand the repeated closures. Although we now live in hope that one of the hastily developed vaccines will bring back our old unthinking freedoms, the prospects for the new year still look very uncertain. For those who've survived it, Covid-19 has left scars on the lungs and on the mind that will be slow to heal, but for those who have lost loved ones, the damage is permanent, and even Christmas will this year be overshadowed by separation and darkness.

One of our culture's most ancient myths is that of Pandora's Box. As well as being among the oldest, it's perhaps the most necessary and valuable story for times like these, since it insists that in a world full of pestilences biological and moral hope somehow persists. This year, so strange and chastening for human beings, may end up being a very good one for birds. Despite reverses in some places, most species have fledged more young than usual, counteracting in however small a way the bigger picture of loss and depletion and, more valuable still, rebutting fatalism by showing that trends can be reversed. And if there's hope for them, then there's also hope for us.

One sign of a renewed connection between people and birds has been this year's surge of interest in birdsong. However it's calculated, in sales of birdsong CDs, downloading of birdsong apps, letters to the editor or bird-related blogposts, this new cultural interest in birdsongs and their singers wouldn't have happened had it not been for the pandemic. So if we ask when the healing is to begin, perhaps, like one season preparing itself silently inside another, it's always already here.

As I emerge from the wood, I think about all the places the birds of my town enriched and made more familiar in the middle months of this year. They renewed the place for me, and I'll not look at it in the same way again. The summer visitors are now long gone, of course, and those that survived the trials of their migration will now be flitting and feeding, and sometimes singing for practice or for the sheer enjoyment of it, in warmer climes. But twice already I've heard the whistles of redwing over-flying the house at twilight, and at low tide the distant shoreline is thickening daily with the winter waders whose cries sometimes carry up to the house through the night air.

And the most familiar birds, though less obtrusive and noisy than they were in spring, remain to share the winter with us. The wrens are still busy in the wood, the thrushes band together to pick berries from the thickets, and the sparrows, now roosting deep in the densest evergreens, still visit gardens and pavements to prospect for scraps. When the weather is poor out at sea, more gulls come to the park lake, and they're sometimes joined by a couple of cormorants into which the bigger fish disappear, still flapping, as if into long, black, upright body-bags. The black-birds' alarm calls and anxious whistles are as numerous now as in summer, but their song is seldom heard. Only on one day recently, abnormally sunny and warm, did I hear a cock bird start up suddenly from a holly bush: a black quill with a gold nib, scrawling his song on the thin grey air of the winter park, like a pledge of spring in winter.

At this turning of the year, material things begin to decay and revert. Nests that once were sticks gradually disintegrate to sticks again, with the occasional jackdaw feather joining them as they clatter down the flue. The boundaries of wood and flower-bed are demilitarized; the summer's song-posts are now anyone's to use. No skylarks sing now above the dunes, from where the autumn gales send waist-high sand-devils racing down to the beach, and on the tide-line itself each day yields a new tangle of driftwood, seaweed and plastic to be picked through by brisk, tiny shorebirds.

Now that the evenings are colder, birds that for most of the year go about singly or in pairs not only increasingly feed communally but also congregate to roost. The smaller the bird, the more vulnerable it is to cold, so wagtails, wrens, goldcrests and long-tailed tits come together in the evening to occupy any hedge, tree crevice or wall cavity not already stuffed with drowsing bats. As for the more naturally gregarious birds, their numbers now swell with newcomers from outside town. The jackdaws – this year's chicks already fast losing their juvenile feathers – gather each evening in a 400-strong flock, and before they bed down in their choice holm oaks they perform an astonishing pre-roost ritual. On no apparent signal they all rise up in great, fast, noisy murmurations, pulsing every which way in oblong torsion, frequently casting off slowcoaches from the centrifuge, who flap madly for readmittance to the clattering, clamouring, shape-shifting ball of birds. The spectacle isn't beautiful but sublime, which is to say that there's something frightening about it, something thrillingly but forbiddingly untameable.

The starlings have left the odd zone between allotment, barracks and prison and are now split between two big roosts. One is at the superstore, where the birds sleep on the roof and occupy most of the day in the thin trees of the car-park or else waddle around, under and between the cars. They're as noisy as ever, and seemingly oblivious of the shoppers as they whistle and ramble with all their life-affirming insouciance.

Much as I love the antics of this town flock, at dusk I prefer to walk down to the shore, where the second group spends the early evenings socializing noisily in the threadbare poplars that grow where the park stream takes a final listless curve before it meets the incoming tide. Out in the bay, red and green lights blink from anchored ships, while closer, near the shore, a few fires glow among the dunes. I'm watching the lights when suddenly the starlings erupt from the trees and wheel around in a dense mass, silent but for the collective hollow rush of their bodies through the evening air heavy with smoke and brine. Then, still silent, they descend in an instant as if they'd collided with an invisible barrier, to settle in the dark reeds like a dream in the mind of a sleeper. Long before I'm awake they'll have lifted off again to probe on the shore or on the sodden playing fields behind the coast road. But for now they must sleep, as I must too. Looking back for a final time to where the starlings went down, I glance beyond the reeds, across the great black absence of the bay, to the distant lights of the far shore. As I turn inland there comes from over the sea the single clipped whistle of a redwing, answered a few seconds later by another, and the sound in the silence is companionable, like a frail, strong light, guiding me home.

Starling

Adlestrop

Yes. I remember Adlestrop —
The name, because one afternoon
Of heat the express-train drew up there
Unwontedly. It was late June.

The steam hissed. Someone cleared his throat.
No one left and no one came
On the bare platform. What I saw
Was Adlestrop — only the name

And willows, willow-herb, and grass,
And meadowsweet, and haycocks dry,
No whit less still and lonely fair
Than the high cloudlets in the sky.

And for that minute a blackbird sang
Close by, and round him, mistier,
Farther and farther, all the birds
Of Oxfordshire and Gloucestershire.

Edward Thomas (1878–1917)

Aderyn du

Aderyn du a'i blufyn sidan,
A'i big aur a'i dafod arian,
A ei di dros ta'i i Gydweli,
I holi hynt yr un'rwy'n garu.

Un, dou, tri pheth sy'n anodd i mi,
Yw cyfri'r ser pan fo hi'n rhewi,
A doti'n llaw i dwtsh a'r lleard,
A deall meddwl f'annwyl gariad.

Blackbird with wings of silk
And golden beak and silver tongue
Will you go for me to Cydweli,
To ask how my love is?

One, two, three things are difficult for me:
Counting the stars when it is freezing
Placing my hand so that I can touch the moon,
And understanding the mind of my loved one.

Trad. Welsh

Sumer is icumen in

Sumer is icumen in,
Lhude sing cuccu!
Groweþ sed and bloweþ med
And springþ þe wde nu,
Sing cuccu!

Awe bleteþ after lomb,
Lhouþ after calue cu.
Bulluc sterteþ, bucke uerteþ,
Murie sing cuccu!
Cuccu, cuccu, wel singes þu cuccu;
Ne swik þu nauer nu.
Sing cuccu nu. Sing cuccu.
Sing cuccu. Sing cuccu nu!

Anon, thirteenth century

Thou bonnie wood o Craigielee

Thou bonnie wood o Craigielee,
Thou bonnie wood o Craigielee,
Near thee I pass'd life's early day,
An won my Mary's heart in thee.

The brume, the brier, the birken bush,
Bloom bonnie o'er thy flowery lee,
An a the sweets that ane can wish
Frae Nature's han, are strewed on thee.
Thou bonnie wood o Craigielee,
Thou bonnie wood o Craigielee,
Near thee I pass'd life's early day,
An won my Mary's heart in thee.

Far ben thy dark green plantin's shade
The cushat croodles am'rously,
The mavis, doon thy buchted glade,
Gars echo ring frae ev'ry tree.
Thou bonnie wood o Craigielee,
Thou bonnie wood o Craigielee,
Near thee I pass'd life's early day,
An won my Mary's heart in thee.

Awa, ye thochtless, murd'rin gang,
Wha tear the nestlin's ere they flee!
They'll sing you yet a cantie sang,
Then, oh! in pity let them be!

Thou bonnie wood o Craigielee,
Thou bonnie wood o Craigielee,
Near thee I pass'd life's early day,
An won my Mary's heart in thee.

Whan winter blaws in sleety showers,
Frae aff the Norlan hills sae hie,
He lichtly skiffs thy bonnie bow'rs,
As laith tae harm a flow'r in thee.
Thou bonnie wood o Craigielee,
Thou bonnie wood o Craigielee,
Near thee I pass'd life's early day,
An won my Mary's heart in thee.

Though fate should drag me south the line,
Or o'er the wide Atlantic sea,
The happy hours I'll ever mind
That I, in youth, hae spent in thee.
Thou bonnie wood o Craigielee,
Thou bonnie wood o Craigielee,
Near thee I pass'd life's early day,
An won my Mary's heart in thee.

Robert Tannahill (1774–1810)

The woodpigeon's nest

Roaming the little path 'neath dotterel trees
Of some old hedge or spinney side I've oft
Been startled pleasantly from musing ways
By frighted dove that suddenly aloft
Sprung through the many boughs with cluttering noise
Till free from such restraints above the head
They smacked their clapping wings for very joys
And in a curious mood I've oft been led
To climb the twig-surrounded trunk and there
On some few bits of stick two white eggs lie
As left by accident – all lorn and bare
Almost without a nest yet bye and bye
Two birds in golden down will leave the shells
And hiss and snap at wind-blown leaves that shake
Around their home where green seclusion dwells
Till fledged, and then the young adventurers take
The old ones' timid flights from oak to oak
Listening the pleasant sutherings of the shade
Nor startled by the woodman's hollow stroke
Till summer's pleasant visions fade
Then they in bolder crowds will sweep and fly
And brave the desert of a winter sky.

John Clare (1793–1864)

A Wind that rose

A Wind that rose
Though not a Leaf
In any Forest stirred
But with itself did cold engage
Beyond the Realm of Bird —
A Wind that woke a lone Delight
Like Separation's Swell
Restored in Arctic Confidence
To the Invisible —

Emily Dickinson (1830–86)

Travelling at dawn to Pa Pass

I set out for Pa Pass at dawn,
heavy with sorrow for the city left behind.
A woman washed in the river's dazzle
and birds in multitude welcomed the sun.
Boats were spread with market wares
beneath my mountain path stitched
by precipitous bridges

Still I climbed.
Unknown settlements showed along
the threads of distant rivers.
I stopped, seized suddenly by loneliness and fear.
That was when an oriole sang clear
and golden, the tune familiar as a friend.
I listened, smiled, walked on.

Wang Wei (699–759)

Acknowledgements

First and greatest thanks to Judit Varga and Michael Malay, who gave me the opportunity and constant encouragement to write. Thanks also to Mihály Bertics and Györgyi Szabó, and to Bath Spa University for producing *On the Wild Side*, an anthology of writing from the MA in Travel and Nature Writing, which featured my work. Tessa David and Richard Atkinson believed in the book from the start – thank you! Finally, special gratitude to Katie Marland for her beautiful artwork.